INTERSCIENCE TRACTS ON PHYSICS AND ASTRONOMY

Edited by R. E. Marshak
University of Rochester

1. D. J. Hughes
   NEUTRON OPTICS
2. M. S. Livingston
   HIGH–ENERGY ACCELERATORS
3. L. Spitzer, Jr.
   PHYSICS OF FULLY IONIZED GASES
   second edition
4. T. G. Cowling
   MAGNETOHYDRODYNAMICS
5. D. ter Haar
   INTRODUCTION TO THE PHYSICS OF MANY-BODY SYSTEMS
6. E. J. Öpik
   PHYSICS OF METEOR FLIGHT IN THE ATMOSPHERE
7. K. Mendelssohn
   CRYOPHYSICS
8. J. L. Delcroix
   INTRODUCTION TO THE THEORY OF IONIZED GASES
9. T. E. Sterne
   AN INTRODUCTION TO CELESTIAL MECHANICS
10. J. Weber
    GENERAL RELATIVITY AND GRAVITATIONAL WAVES
11. R. E. Marshak and E. C. G. Sudarshan
    INTRODUCTION TO ELEMENTARY PARTICLE PHYSICS
12. J. L. Olsen
    ELECTRON TRANSPORT IN METALS

13. M. Francon
    MODERN APPLICATIONS OF PHYSICAL OPTICS

14. P. B. Jones
    THE OPTICAL MODEL IN NUCLEAR AND PARTICLE PHYSICS

15. R. K. Adair and E. C. Fowler
    STRANGE PARTICLES

16. R. Wilson
    THE NUCLEON–NUCLEON INTERACTION: EXPERIMENTAL AND PHENOMENOLOGICAL ASPECTS

17. J. F. Denisse and J. L. Decroix
    PLASMA WAVES

18. W. F. Brown, Jr.
    MICROMAGNETICS

19. A. Rose
    CONCEPTS IN PHOTOCONDUCTIVITY AND ALLIED PROBLEMS

20. A. Guinier and D. L. Dexter
    X-RAY STUDIES OF MATERIALS

21. T. G. Northrup
    THE ADIABATIC MOTION OF CHARGED PARTICLES

22. G. Barton
    INTRODUCTION TO ADVANCED FIELD THEORY

23. C. D. Jeffries
    DYNAMIC NUCLEAR ORIENTATION

24. G. B. Benedek
    MAGNETIC RESONANCE AT HIGH PRESSURE

25. D. L. Dexter and R. S. Knox
    EXCITONS

26. D. H. Sampson
    RADIATIVE CONTRIBUTIONS TO ENERGY AND MOMENTUM TRANSPORT IN A GAS

27. P. Carruthers
    INTRODUCTION TO UNITARY SYMMETRY

28. L. Spitzer, Jr.
    DIFFUSE MATTER IN SPACE

Additional volumes in preparation

M16, an H II region in our Galaxy. This bright nebula, surrounding an O5 star, is also known as NGC 6611. Photograph obtained in red light with the Palomar 200-inch telescope, Mt. Wilson and Palomar Observatories.

DIFFUSE MATTER IN SPACE

LYMAN SPITZER, JR.
Princeton University Observatory
Princeton, New Jersey

INTERSCIENCE PUBLISHERS
a division of John Wiley & Sons, New York • London • Sydney • Toronto

Copyright © 1968 by John Wiley & Sons, Inc.

All rights Reserved. No part of this book may be reproduced by any means, nor transmitted, nor translated into a machine language without the written permission of the publisher.

Library of Congress Catalog Card Number 68-29397
SBN 470 817100
Printed in the United States of America

To my wife

Preface

The tenuous matter between stars is of interest in several ways. The interstellar material plays a very major part in the evolution of stars and of the Galaxy; groups of new stars are believed to have their birthplace within interstellar clouds, and ejection of matter from old stars, especially from supernovae, enriches the interstellar gas in its abundance of heavy elements. In addition, the diverse physical processes which occur between the stars are in many ways quite different from those which are important in normal stellar atmospheres or deep within the stars; the enormous deviations from thermal equilibrium, the extended time scale for dynamical processes—much longer than the time scale for radiative cooling—and the complex interactions between the energetic ions, the relatively cold gas, and the magnetic field are examples of the intricate physical processes which make interstellar matter research both challenging and fascinating.

This brief text is designed to present in a unified way the general theoretical basis for this research. Written for graduate students with a good general background in both physics and astronomy, its scope is limited to what might reasonably be covered within a single course. The treatment is, therefore, very brief; many important subjects are omitted entirely, or are mentioned only in passing. Preference has been given to those physical processes which have been sufficiently analyzed so that their significance in the interstellar medium seems reasonably well established. As a result, some of the most exciting but also rather speculative topics in modern astrophysics, such as the origin of spiral arms and physical conditions in the galactic corona and in intergalactic space, are not discussed, while the expansion of gas around newly born hot stars or around supernovae has been treated at some length. The discussion is generally rather condensed; a fuller presentation of all intermediate steps would make the text easier to read, but might not increase its educational value.

The first three chapters of the book provide the background necessary for the remaining three, more theoretical, chapters. In the first chapter appears a brief historical introduction, together with a description of the gravitational limit on the local density of interstellar material. The primary observational results on which the subsequent theory must be based are summarized in Chapters 2 and 3 in a very abbreviated form; the primary emphasis is on how the observational data, obtained with optical or radio telescopes, are interpreted in terms of the number and velocity distribution of the absorbing or emitting particles—atoms, ions, molecules, or solid grains. Problems relating to how the observations are actually obtained have been discussed by astronomers more familiar with this fundamental aspect of the subject, and are not discussed here. Chapter 4 treats the physical processes which are associated with encounters between various types of interstellar particles and which determine, for example, the ionization level and the velocity distribution of the atoms, and the size, composition, and orientation of the grains. A brief introduction to problems of interstellar gas dynamics appears in Chapter 5. Some of the various phenomena that may be important in a future theory of star formation are discussed in Chapter 6.

The references at the end of each chapter are in no sense complete, but are designed to help the reader gain additional information if desired. When adequate surveys of a subject are available, reference is frequently made to these surveys rather than to individual papers. Hence the list of references does not indicate fairly who have been responsible for which results, and for this reason virtually no names are mentioned in the text except in the historical introduction.

Many astronomers have been helpful in the preparation of this manuscript, including R. D. Davies, G. R. A. Ellis, H. L. Johnson, F. D. Kahn, J. H. Oort, S. R. Pottasch, V. C. Reddish, M. J. Seaton, and F. J. Smith. Particularly important suggestions were made by J. M. Greenberg and K. Serkowski on Chapter 3, by

PREFACE

A. Dalgarno and D. E. Osterbrock on Chapter 4, and by L. Mestel on Chapter 6. The extensive comments and corrections made by the graduate students at Princeton have been highly valuable; in particular, M. G. Tomasko has checked several of the chapters in detail. I am particularly indebted to J. P. Ostriker for a critical reading of the entire manuscript and to J. B. Rogerson for his collaboration at a much earlier stage some 15 years ago, in the preparation of a first draft for what was then planned as a much more ambitious monograph on what was at that time a less extensive subject.

This book was largely written during the period September 1966 through January 1967, while the author was at the Institut d'Astrophysique as Professeur Associé of the University of Paris. With a few exceptions the material covered here is necessarily restricted to results available before 1967. It is a pleasure to express here my appreciation for the appointment in Paris that made this work possible.

Lyman Spitzer, Jr.

Princeton University Observatory
September, 1968

Contents

1. **Introduction** 1
   - 1.1 Historical Development 1
   - 1.2 Gravitational Effects of Interstellar Matter . . . 4

2. **Observations of the Gas** 9
   - 2.1 Theory of Emission and Absorption by Gas . . . 9
     - Basic concepts 9
     - Excitation and ionization 11
     - Line emission and absorption 14
     - Line profile and curve of growth 17
     - Continuous emission and absorption by hydrogenic atoms 21
     - Synchrotron emission 22
   - 2.2 Emission from the Gas 24
     - Hydrogen 21-cm line 24
     - Visual emission lines 29
     - Radio continuum and line emission 33
     - OH emission 35
   - 2.3 Absorption by the Gas 36
     - Radio lines 37
     - Visual lines 39
   - 2.4 Nonthermal Emission 44
     - Observed brightness temperature 45
     - Energetic particles 47
     - Interstellar magnetic field 49

3. **Observations of the Grains** 57
   - 3.1 Theory of Absorption and Scattering by Grains . . 57
     - Spheres 59
     - Cylinders and spheroids 62
     - Complex particles 64

3.2	Extinction and Scattering by the Grains		65
	Selective extinction		65
	General extinction		69
	Scattering		71
3.3	Polarization		72
	Dependence on color excess		73
	Dependence on wavelength		75
	Dependence on galactic longitude		76
3.4	Spatial Distribution of Interstellar Matter		79
	Visible nebulae		80
	Indirect and statistical evidence		82
4.	**Interactions among Interstellar Particles**		**88**
4.1	Collisional Processes		88
	Elastic collisions, short-range forces		90
	Elastic collisions, electrostatic forces		94
	Excitation by electron impact		95
	Excitation by atoms		98
	Recombination and ionization		99
4.2	Excitation and Kinetic Equilibrium		101
	Velocity distribution function		103
	Excitation		106
4.3	Ionization and Dissociation		109
	Hydrogen ionization		112
	Ionization of sodium and calcium		120
	Dissociation equilibrium for molecules		125
4.4	Kinetic Temperature		126
	H II region of pure hydrogen		129
	H II region with impurities		133
	H I region		136
	Equipartition of kinetic energy		140
	Thermal instability		141
4.5	Equilibrium Properties of the Grains		142
	Temperature of the solid material		142
	Electric charge		145
	Orientation		147
4.6	Evolution of Grains and Formation of Molecules		152
	Growth of grains		152
	Disruption of grains		154
	Molecule formation		157

CONTENTS

5. Dynamics of the Interstellar Gas 161
   5.1 Dynamical Principles and Problems 161
        Basic equations 162
        Virial theorem 164
        Shock fronts 166
        Parameters of the interstellar medium . . . 170
        Energy balance in cloud collisions 172
   5.2 Equilibrium Density Distribution 175
        Galactic distribution 176
        Distribution perpendicular to the galactic plane . 179
        Distribution in the galactic disc 181
   5.3 Expansion of H II Regions 183
        Ionization fronts 184
        Initial ionization of the gas 187
        Expansion of the ionized gas 188
        Efficiency of acceleration 193
   5.4 Supernova Shells 194
        Initial expansion of a supernova atmosphere . . 195
        Intermediate nonradiative expansion 198
        Late isothermal expansion 200
   5.5 Interactions between Clouds and Stars . . . 202
        Ionization of an H I cloud in an H II region . . 202
        Acceleration of an H I cloud by the rocket effect . 205
        Radiation pressure on grains near a bright star . 207
        Radiation pressure of galactic light on grains . . 210

6. Formation of Stars 214
   6.1 Gravitational Instability 214
        Linearized perturbation equations 215
        Instability of a gaseous disc 217
        Collapse of an isolated cloud 220
   6.2 Fragmentation of a Collapsing Cloud 225
        Uniform nonrotating cloud, $\mathbf{B} = 0$ 225
        Rotating cloud, $\mathbf{B} = 0$ 230
        Uniformly magnetized cloud, $\Omega = 0$ 232
   6.3 Later Stages in Star Formation 234
        Radiative decrease of energy 235
        Decrease in magnetic flux 238
        Decrease in angular momentum 242

Symbols 247

Index 255

Chapter 1

Introduction

1.1 Historical Development

The concept of a pervasive interstellar gas, extending throughout our Galaxy, did not gain general acceptance until after the first quarter of the twentieth century. As early as 1904 J. Hartmann [1] had observed stationary absorption lines of ionized calcium in a spectroscopic binary, but the evidence at that time was insufficient to verify his suggestion that the lines might be interstellar in origin rather than circumstellar; in 1922 the evidence available permitted the conclusion [2], "The hypothesis that the calcium cloud is a part of the star's atmosphere seems the most satisfactory."

In the subsequent decade progress was rapid both in observation and in theory. A number of workers found that the absorption lines of ionized calcium in stellar spectra of types B and O increased in intensity with distance; in addition, the radial velocities of these lines showed the double sine wave characteristic of galactic rotation, but with an amplitude corresponding to half the stellar distance, as would be expected for a uniform gas extending from the Sun to the early-type star. In particular, the extensive data assembled on these two aspects by Plaskett and Pearce [3]—who give also a full historical account of earlier work—showed conclusively that the interstellar calcium gas must extend more or less uniformly throughout space. Eddington [4] in his celebrated Baker Lecture laid the foundation for much of the modern theory of the interstellar gas, computing both the kinetic temperature and the ionization level of the atoms assumed to be present.

It was at about this same time that the concept of a general absorbing medium in the Galaxy, composed of small solid particles, or dust grains, also began to gain general acceptance. While the presence of such an absorbing layer had been deduced from the star counts as early as 1847 by F. G. W. Struve [5], whose mean extinction coefficient* of 1 mag/kpc is not far below present estimates, as late as 1923 the evidence for either general or selective extinction was not really conclusive [6]. Reliable evidence for general extinction required objects of known brightness at a distance determined geometrically rather than photometrically. These were provided in 1930 by Trumpler's measures [7] of the diameters of galactic clusters and Bottlinger's and Schneller's investigation [8] of the distribution of Cepheids perpendicular to the galactic plane; van de Kamp's [9] analysis of the concentration of external galaxies to the poles of our own Galaxy provided yet one more relatively conclusive argument. By 1935 a layer of dust as well as of gas extending throughout the Galaxy was generally accepted.

Since the 1930's our picture of the interstellar medium has been further influenced by two major developments in observational astronomy. First there has been increasing evidence that the distribution of both gas and dust is far from uniform. From their pioneering photoelectric measurements of selective extinction in B and O stars Stebbins, Huffer and Whitford [10] concluded, "The absorption in the Galaxy is obviously so irregular and spotted that a constant coefficient of absorption cannot be used for any large region of space." The complex interstellar absorption

---

* In the present work extinction will be used to denote the sum of scattering and true absorption by the solid interstellar particles (see Chapter 3).

1. INTRODUCTION

line profiles obtained by Adams [11] in 1949, following earlier work by Beals [12] and by Merrill and Wilson [13], showed that the interstellar gas has quite different radial velocities in different regions. These observations led to the concept that much of the interstellar medium is concentrated in irregular clouds and cloud complexes, a point of view which is generally consistent with the cloudy structure evident in photographs of the Milky Way and of most nebulae (see the frontispiece). According to this point of view, which has been developed by Ambartsumian [14] and his co-workers, the bright diffuse nebulae, which have been known for many years, are simply the clouds which happen to be sufficiently close to bright stars to shine either by reflection or by fluorescence, while the dark nebulae are those which happen to be projected against a luminous background.

A second major development has been the extensive observational material on the radiation by neutral hydrogen at 21-cm wavelength, following the theoretical prediction of such interstellar radiation by van de Hulst [15] in 1945. The concentration of hydrogen to spiral arms and the delineation of these all across our Galaxy are fundamental results of these observations. Our knowledge of the density, temperature, and velocity distribution of the hydrogen gas, while by no means complete, has been much enhanced by these radio-astronomical studies.

On the theoretical side two developments since 1935 have had special importance for the modern era of interstellar matter research. One of these is Strömgren's investigation [16] of the regions surrounding early-type stars, showing that the hydrogen is nearly fully ionized inside a sharply bounded volume, and is almost completely neutral outside. The various physical differences and interactions between H II and H I regions, as they are now called, is one of the principal subjects discussed in this book. A picture of an H II region (M16) around an O5 star is reproduced in the frontispiece, a photograph obtained in red light with the 200-inch Hale telescope on Mt. Palomar. This bright nebula, also known as NGC 6611, is illuminated by two early-type stars (HD 168075 and 168076), of spectral types O7 and O5.

Another theoretical development of major significance has been the gradual realization that magnetic fields and energetic, or suprathermal, particles are a major constituent of the interstellar environment. Not only do relativistic electrons account directly for much of the nonthermal radio radiation from the Galaxy, but cosmic-ray protons and heavy ions, and the magnetic field which deflects them, must play a major role in the equilibrium and acceleration of the interstellar medium. Our understanding of these important effects is still in its infancy.

1.2 Gravitational Effects of Interstellar Matter

Many possible forms of interstellar matter, such as highly ionized atoms, many types of molecules, and any solid objects much bigger than a few microns in size, would be difficult to detect directly from radiation reaching the Earth, even if the total mass in these forms were much greater than in the form of neutral hydrogen atoms. In discussing physical processes in interstellar space it is very helpful, therefore, to have a limit on the total mass of material present between the stars. Such a limit can be obtained from the total gravitational attraction of the matter within a few hundred parsecs of the Sun. Since our Galaxy is a flattened disc, measurement of the gravitational acceleration, g_z, perpendicular to the disc gives directly the total mass of the material in the solar neighborhood, including both interstellar matter and stars. In particular, g_z is determined by measuring the density gradient in the z direction of some group of stars whose velocities in the z direction can also be measured.

This section gives the theory which relates the measured density gradient to g_z. Since a group of stars does not constitute a fluid in the usual sense, it is not obvious that the usual equations of a fluid in equilibrium are applicable, and instead Liouville's Theorem may be used. This theorem, which is applicable to an assembly of mass points moving in a potential field, with no dissipation, states that f, the density of points in phase space, is constant along a dynamical trajectory. It is assumed here that the Galaxy is

1. INTRODUCTION

stratified in plane parallel layers, and is in a statistically steady state; these two assumptions should provide a reasonable first approximation. We may ignore the momenta p_x and p_y in the x and y directions, and all quantities are functions of z and p_z only; $f(p_z, z)\, dp_z\, dV$ gives the number of stars within the interval dp_z and within the volume element dV. In this simple one-dimensional case the motion in z is completely described by the energy integral, E, where

$$E = \frac{p_z^2}{2m} + m\phi(z); \qquad (1\text{-}1)$$

m is the mass of a star, assumed to be the same for all stars in the group, and $\phi(z)$ is the gravitational potential, a function of height, z, above the midplane of the Galaxy. For stars of each value of E, $f(p_z, z)$ will be constant along a dynamical trajectory and must therefore be a function of E only, which we write as $f(E)$.

The number density, n, of stars per unit volume is related to f by the equation

$$n = \int_{-\infty}^{+\infty} f(E)\, dp_z \qquad (1\text{-}2)$$

while the mean square momentum in the z direction is given by

$$\langle p_z^2 \rangle = \frac{1}{n} \int_{-\infty}^{+\infty} f(E) p_z^2 \, dp_z \qquad (1\text{-}3)$$

If we multiply equation (1-3) by n and differentiate with respect to z, we obtain

$$\frac{d}{dz}(n\langle p_z^2 \rangle) = -mg_z \int_{-\infty}^{+\infty} \frac{df(E)}{dE} p_z^2 \, dp_z \qquad (1\text{-}4)$$

where we have differentiated under the integral sign and used the relation

$$\frac{\partial E}{\partial z} = m\frac{d\phi}{dz} = -mg_z \qquad (1\text{-}5)$$

Since also from equation (1-1)

$$\frac{df(E)}{dE} = \frac{m}{p_z}\frac{\partial f}{\partial p_z} \tag{1-6}$$

we can integrate equation (1-4) by parts and obtain, substituting the velocity w_z for p_z/m,

$$\frac{1}{n}\frac{d}{dz}(n\langle w_z^2\rangle) = g_z \tag{1-7}$$

The total mass density, ρ, at any point is obtained from Poisson's equation, which gives, using equation (1-5),

$$\rho(z) = -\frac{1}{4\pi G}\frac{dg_z(z)}{dz} \tag{1-8}$$

Equation (1-7) has been applied to K giants [17], which form a reasonably homogeneous group and are sufficiently bright and numerous to provide good statistics. The resultant values of g_z are plotted in Figure 1.1. From equation (1-8) the total mass density at the midplane is found to be 10.0×10^{-24} g/cm^3. The mass density of known stars is about 4×10^{-24} g/cm^3. Since the compilation of stars may be incomplete at the faint end, we have for

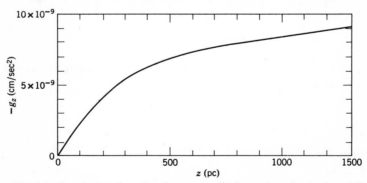

Fig. 1.1. Gravitational acceleration perpendicular to the galactic plane. The curve, taken from ref. [17], shows the measured values of $-g_z$, in cm/sec^2, as a function of z, the height above the galactic plane, in parsecs.

1. INTRODUCTION

ρ_{Int}, the total density of interstellar matter in the solar neighborhood

$$\rho_{\text{Int}} \leq 6.0 \times 10^{-24} \text{ g/cm}^3 \qquad (1\text{-}9)$$

corresponding to 2.6 H atoms per cm^3, if a He/H ratio of 10% by number is assumed. This upper limit on ρ_{Int} is sometimes known as the "Oort limit." When account is taken of the small thickness of the interstellar material—about 200 pc perpendicular to the galactic plane—and the uncertainty of the observations plotted in Figure 1.1, a value of ρ_{Int} as great as 10×10^{-24} g/cm^3 would not be entirely excluded [17,18]. The gravitational limit on ρ_{Int} is of fundamental importance in the study of the interstellar medium.

Similar analyses, in principle, have been made for other galaxies, and for other regions in our own Galaxy, and the gravitational mass has then been compared with the observed stellar luminosity. In some systems, particularly in clusters of elliptical galaxies, the ratio of mass to luminosity much exceeds its value for stars near the sun; the excess mass could conceivably be due to some form of interstellar (or intergalactic) matter. However, the stellar content of these remote systems is so uncertain that assigning the measured gravitational mass to various stellar and interstellar components is highly uncertain.

References

1. J. Hartmann, *Sitzb. Kgl. Akad. Wiss.*, 527 (1904) translated in *Astrophys. J.*, **19**, 268 (1904).
2. R. K. Young, *Publ. Dominion Astrophys. Obs. Victoria, B.C.*, **1**, 219 (1922).
3. J. S. Plaskett and J. A. Pearce, *Publ. Dominion Astrophys. Obs. Victoria, B.C.*, **5**, 167 (1933).
4. A. S. Eddington, *Proc. Roy. Soc. (London), Ser. A*, **111**, 424 (1926).
5. F. G. W. Struve, *Etudes d'astronomie stellaire*, Académie Impériale des Sciences, St. Petersburg, 1847.
6. H. Kienle, *Jahrb. Radioactivität und Elektronik*, **20–1**, 46 (1923).
7. R. J. Trumpler, *Lick Obs. Bull.*, **14**, 154 (1930); No. 420.
8. K. F. Bottlinger and H. Schneller, *Z. Astrophys.* **1**, 339 (1930).
9. P. van de Kamp, *Astronom. J.*, **42**, 97 (1932).

10. J. Stebbins, C. M. Huffer, and A. Whitford, *Astrophys. J.*, **91**, 20 (1940).
11. W. S. Adams, *Astrophys. J.*, **109**, 354 (1948).
12. C. S. Beals, *Monthly Notices, Roy. Astron. Soc.*, **96**, 661 (1936).
13. P. W. Merrill and O. C. Wilson, *Astrophys. J.*, **86**, 44 (1937).
14. V. A. Ambartsumian and S. G. Gordeladse, *Bull. Abastumani*, **2**, 37 (1938).
15. H. C. van de Hulst, *Ned. Tijdschr. Natuurk.*, **11**, 201 (1945).
16. B. Strömgren, *Astrophys. J.*, **89**, 526 (1939).
17. J. H. Oort, in *Stars and Stellar Systems*, Vol. 5, Univ. of Chicago Press, Chicago, 1965, p. 455.
18. R. J. Gould, T. Gold and E. E. Salpeter, *Astrophys. J.*, **138**, 414 (1963).

Chapter 2

Observations of the Gas

2.1 Theory of Emission and Absorption by Gas

All our information on the material between the stars is based on measurements either of electromagnetic radiation or of cosmic-ray particles. The present chapter summarizes briefly the observational facts obtained about interstellar electrons, atoms, and molecules, either by measurement of the electromagnetic radiation emitted or absorbed by such particles, or by measurement of the number of energetic electrons and stripped atomic nuclei reaching the Earth as cosmic rays. Knowledge obtained from radiation which has interacted with the solid particles in interstellar space is considered in the next chapter. To understand how the observed photon intensities have been interpreted in terms of the density, kinetic temperature, composition, and velocity of the interstellar gas requires some understanding of the interaction between electrons, atoms, and molecules on the one hand, and photons on the other. This first section is devoted to a summarized theory of these interactions. Standard texts in quantum theory and statistical mechanics should be consulted for the derivation from first principles of the basic equations given here.

Basic Concepts

The photons traveling by a point $\mathbf{r}$ at a time t will each have a different direction, $\mathbf{\kappa}$, and a different value of the frequency, v.

To characterize the radiation field we must specify the energy passing by as a function of all four of these physical variables. We define the specific intensity $I_\nu(\kappa, \mathbf{r}, t)$ so that $I_\nu\, d\nu\, d\omega\, \mathbf{dA}\, dt$ is the energy of those particles which during a time interval dt pass through the area $\mathbf{dA}$, whose frequency lies within the element $d\nu$ about ν, and whose direction is within the solid angle $d\omega$ about κ; the area $\mathbf{dA}$ is located at the position $\mathbf{r}$ and is perpendicular to the photon direction, κ. The change of I_ν resulting from interaction with matter is governed by the equation of transfer. This equation is derived by considering the flow of energy in and out the ends of a cylinder of length ds, with the use of the absorption and emission coefficients, κ_ν and j_ν. The emission coefficient, or emissivity, $j_\nu(\kappa, \mathbf{r}, t)$, is defined so that $j_\nu\, dV\, d\nu\, d\omega\, dt$ is the energy emitted by the volume element dV (equal to $ds\, dA$) in the intervals $d\nu$, $d\omega$, and dt, while $\kappa_\nu I_\nu\, dV\, d\nu\, d\omega\, dt$ is the corresponding energy absorbed from a beam of specific intensity I_ν. If we assume that the photons travel in straight lines, the change of I_ν along a distance ds, taken along a light ray, then equals [1,2]

$$\frac{dI_\nu}{ds} = -\kappa_\nu I_\nu + j_\nu \tag{2-1}$$

If we define the "optical thickness" τ_ν along the ray path by the expression

$$d\tau_\nu = \kappa_\nu\, ds \tag{2-2}$$

and consider the radiation received from a region of total optical thickness τ_ν, equation (2-1) may be integrated to yield

$$I_\nu = I_\nu(0)e^{-\tau_\nu} + \int_0^{\tau_\nu} \frac{j_\nu}{\kappa_\nu} e^{-(\tau_\nu - \tau_\nu')}\, d\tau_\nu' \tag{2-3}$$

$I_\nu(0)$ denotes the intensity of radiation incident on the far side of the region, where τ_ν is taken to be zero; j_ν/κ_ν may vary with τ_ν'.

Under some conditions j_ν and κ_ν have the same relative values as they would in strict thermodynamic equilibrium, where I_ν is given by the familiar Planck function

$$B_\nu(T) = \frac{2h\nu^3}{c^2} \frac{1}{e^{h\nu/kT} - 1} \tag{2-4}$$

2. OBSERVATIONS OF THE GAS

Since dI_v/ds must vanish if I_v is constant, we see from equation (2-1) that the ratio j_v/κ_v in thermodynamic equilibrium must be given by

$$j_v = \kappa_v B_v(T) \tag{2-5}$$

a relationship known as Kirchhoff's law.

If equation (2-5) is assumed to hold in equation (2-1), and if the temperature T appearing in j_v/κ_v is constant over the path length, equation (2-3) becomes

$$I_v = I_v(0)e^{-\tau_v} + B_v(T)\{1 - e^{-\tau_v}\} \tag{2-6}$$

For measurements in the radio spectrum I_v is frequently replaced by the "brightness temperature," T_b, defined as the temperature at which $B_v(T)$ equals the observed I_v. Since hv/kT is generally very much less than unity for all radio frequencies, we may expand the exponent in equation (2-4), neglecting terms of second and higher order in hv/kT, obtaining

$$I_v = \frac{2v^2 k T_b}{c^2} \tag{2-7}$$

The quantity T_b has the advantage that it involves only one unit (degrees) instead of five (energy, time, area, band width, and solid angle). In terms of T_b equation (2-6) takes the simple form

$$T_b = T_{b0} e^{-\tau_v} + T(1 - e^{-\tau_v}) \tag{2-8}$$

where T_{b0} is the brightness temperature of the incident radiation on the far side of the emitting and absorbing region.

Excitation and Ionization

The volume emission and absorption coefficients, j_v and κ_v, depend on the population of the atomic levels responsible for emitting and absorbing radiation. The detailed discussion of excitation and ionization under interstellar conditions in general requires information on electron–atom collisions, which are treated in Section 4.1. Here we give the relations which are valid in thermodynamic equilibrium, and which are sometimes applicable to the interstellar gas.

Let $n(X_r)$ be the total number of atoms of some element, X, in the stage of ionization r, per cm^3. For a specific atom we shall generally replace the subscript r by a Roman numeral following the name of the element; e.g., n(H I) and n(Ca II) will denote the particle density of neutral hydrogen and singly ionized calcium atoms, respectively, thus applying to the atoms themselves the notation generally employed to denote their spectra. The symbol n_X will denote the particle density of atoms of element X, including all stages of ionization. In addition, let $n_j(X_r)$ be the corresponding number of these atoms which are in the stage of ionization r, and in a particular energy level j, characterized by an energy E_{rj}, and a statistical weight g_{rj}. Then in thermodynamic equilibrium the relative population of two levels, denoted by j and k, is given by the Boltzmann equation

$$\frac{n_j^*(X_r)}{n_k^*(X_r)} = \frac{g_{rj}}{g_{rk}} e^{-(E_{rj}-E_{rk})/kT} \qquad (2\text{-}9)$$

where we use asterisks to denote values in thermodynamic equilibrium. If we take the sum of n_k^* over all k, then the fraction of X atoms in the stage of ionization r which are excited to level j becomes

$$\frac{n_j^*(X_r)}{n^*(X_r)} = \frac{g_{rj} e^{-E_{rj}/kT}}{f_r} \qquad (2\text{-}10)$$

where f_r, the partition function, is defined by

$$f_r = \sum_k g_{rk} e^{-E_{rk}/kT} \qquad (2\text{-}11)$$

The partition of atoms over different stages of ionization is given in thermodynamic equilibrium by the Saha equation

$$\frac{n^*(X_{r+1})n_e}{n^*(X_r)} = \frac{f_{r+1}}{f_r} f_e \qquad (2\text{-}12)$$

2. OBSERVATIONS OF THE GAS

where the partition functions f_r and f_{r+1} are defined in equation (2-11), and where f_e, the partition function of the free electrons per unit electron density, may be shown to be

$$f_e = 2\left(\frac{2\pi mkT}{h^2}\right)^{3/2} = 4.83 \times 10^{15}\, T^{3/2} \quad (2\text{-}13)$$

where the various symbols have their usual meaning—see the list of symbols at the end of the book.

If the partition functions f_r and f_{r+1} are approximated by their first terms, with the ground levels denoted by a subscript 1, the Saha equation takes the more familiar form

$$\frac{n^*(X_{r+1})n_e}{n^*(X_r)} = 2\frac{g_{r+1,\,1}}{g_{r,\,1}}\left(\frac{2\pi mkT}{h^2}\right)^{3/2} e^{-(E_{r+1,\,1} - E_{r,\,1})/kT} \quad (2\text{-}14)$$

As we shall see in Section 4.2, the actual velocity distribution of ions and electrons in the interstellar gas normally follows the Maxwell-Boltzmann law at some kinetic temperature T. Thus collisions between ions and electrons will tend to be identical with those in thermodynamic equilibrium, with the same ion and electron densities and the same T. For some purposes it is convenient to compare the actual value of n_j with its value, n_j^*, in what we shall call the "equivalent thermodynamic system"; i.e., a system in thermodynamic equilibrium in which the temperature equals the actual kinetic temperature, in which the electron density, n_e, is equal to its actual value, and in which the density of atoms of type X ionized $r+1$ times, denoted by $n^*(X_{r+1})$, equals the actual value of $n(X_{r+1})$. Additional positive ions of other elements are assumed present, if necessary, so that electrical neutrality is maintained. The total density of atoms of species X, including all stages of ionization, will generally depend on the stage of ionization $r+1$ used in defining the equivalent thermodynamic system, and will also differ, in general, from the actual interstellar value of n_X. For any level, j, in an atom in stage of ionization r, we define

$$\frac{n_j}{n_j^*} \equiv b_j \quad (2\text{-}15)$$

In equation (2-15) the dependence on stage of ionization is not explicitly indicated, as will be done in all cases where no confusion results from this simplification. Since b_j equals unity for free electrons with a Maxwellian distribution, physical continuity leads one to expect that for bound states approaching very close to the ionization limit b_j should approach unity.

Line Emission and Absorption

We consider first the emission of radiation by an atom with a lower level j and an upper level k; the stage of ionization is again denoted by r. The Einstein probability coefficient for spontaneous downwards transitions from k to j per unit time is denoted by A_{kj}; B_{kj} and B_{jk} are the corresponding probability coefficients for induced downwards and upwards transitions, defined in terms of radiation energy density, U_ν, measured in ergs per cm^3 per frequency interval. If, as before, we let n_{rj}, n_{rk} be the particle densities of atoms in these two levels, then the total energy radiated spontaneously per cm^3 per unit solid angle per second will be given by

$$\int j_\nu \, d\nu = \frac{h\nu_{jk} n_{rk} A_{kj}}{4\pi} \quad (2\text{-}16)$$

where ν_{jk} is the frequency at the line center and the integral extends over the line. The population density of emitting atoms, n_{rk}, may be expressed in terms of $n_e n(X_{r+1})$ by means of equations (2-10), (2-12), and (2-15), equating $n(X_{r+1})$ to $n^*(X_{r+1})$ in the equivalent thermodynamic system. We obtain

$$n_k(X_r) = \frac{b_k g_{rk} e^{-E_{rk}/kT} n_e n(X_{r+1})}{f_{r+1} f_e} \quad (2\text{-}17)$$

Equation (2-17) is useful when most of the atoms of type X are in the stage of ionization $r+1$ and $n(X_{r+1})$ is nearly equal to n_X. For hydrogen atoms $n(X_{r+1})$, also denoted by $n(\text{H II})$, becomes the proton density, n_p; in addition f_2 degenerates to a single numerical factor, since the ionized H atom has no excited states. If we take

2. OBSERVATIONS OF THE GAS

the zero energy to correspond to a free electron at rest, f_2 equals unity for hydrogen, and the total emission in a particular hydrogen line emitted from the level k per cm^3 per sec per unit solid angle becomes

$$\int j_\nu \, d\nu = \frac{b_k g_k h\nu_{jk} A_{kj} n_e n_p e^{-E_k/kT}}{4\pi f_e} \quad (2\text{-}18)$$

where f_e is again the electron partition function, defined in equation (2-13), and the integral again extends over the line; E_k is the binding energy of the level with total quantum number n, and equals $-E_1/n^2$. We have dropped the subscript r from g_k, which equals $2n^2$ for neutral H.

The total absorption of radiation by an atom may be analyzed in a similar way, computing the integral of $\kappa_\nu I_\nu$ over all $d\nu$. This integral, which gives the total absorbed energy in the line per unit solid angle per cm^3 per sec, is given not simply by the stimulated absorption, but rather by the difference between stimulated absorption and stimulated emission, since the photons emitted by stimulated emission are traveling in the same direction as the stimulating photons and have the same frequency. If we assume that I_ν is relatively constant over the line center, then this quantity may be taken out of the integral over $d\nu$, and we obtain for the rate of energy absorption in the line

$$I_{\nu_{jk}} \int \kappa_\nu \, d\nu = h\nu_{jk}(n_j B_{jk} - n_k B_{kj}) I_{\nu_{jk}}/c \quad (2\text{-}19)$$

To simplify equation (2-19) we express κ_ν in terms of an atomic absorption cross section, s_ν, defined by

$$n_j s_\nu = \kappa_\nu \quad (2\text{-}20)$$

We also define an integrated atomic cross section, s, by the relation

$$s = \int s_\nu \, d\nu \quad (2\text{-}21)$$

where the integral again extends over the line. The condition that the right-hand sides of (2-16) and (2-19) must be equal when $B_\nu(T)$

in equation (2-4) is substituted for I_ν gives the familiar relations between the Einstein coefficients

$$g_j B_{jk} = g_k B_{kj} = \frac{c^3}{8\pi h \nu^3} g_k A_{kj} \qquad (2\text{-}22)$$

If now we substitute equations (2-20) and (2-19) into (2-21), substituting from equations (2-10) and (2-15) for n_j and n_k, we obtain

$$s = s_u \left(1 - \frac{b_k}{b_j} e^{-h\nu/kT}\right) \qquad (2\text{-}23)$$

where s_u the value of s uncorrected for stimulated emission, equals

$$s_u = \frac{h\nu_{jk} B_{jk}}{c} \qquad (2\text{-}24)$$

In terms of the usual oscillator strength, f_{jk}, equation (2-24) may be written

$$s_u = \frac{\pi e^2}{mc} f_{jk} = 2.65 \times 10^{-2} f_{jk} \qquad (2\text{-}25)$$

If the downwards oscillator strength is defined by the relation

$$f_{kj} = -\frac{g_j}{g_k} f_{jk} \qquad (2\text{-}26)$$

then from any level j the sum of all the oscillator strengths obeys the sum rule

$$\sum_k f_{jk} = \text{No. of jumping electrons in level } j \qquad (2\text{-}27)$$

If $h\nu/kT$ is large compared with unity, as it usually is throughout most of the visible spectrum, the exponential term in equation (2-23) may be ignored; stimulated emission under these conditions in negligible. If, on the other hand, $h\nu/kT$ is much less than unity, stimulated emission is comparable with the absorption if b_k/b_j is about unity. We may expand the exponent in this case, obtaining

$$s = \frac{\pi e^2 f_{jk}}{mc} \frac{h\nu}{kT} \left\{\frac{b_k}{b_j} + \frac{kT}{h\nu}\left(1 - \frac{b_k}{b_j}\right)\right\} \qquad (2\text{-}28)$$

2. OBSERVATIONS OF THE GAS

If b_k/b_j is unity the quantity in brackets in this equation is unity, and the stimulated emission reduces s below s_u by a factor $h\nu/kT$, which is of order 10^{-4} for transitions in the radio spectrum. If $b_k - b_j$ exceeds $b_k h\nu/kT$, the absorption coefficient becomes negative, producing maser amplification. Evidently for transitions with an energy change much less than kT, s is very sensitive to small deviations of the relative populations from their values in thermodynamic equilibrium.

Line Profile and Curve of Growth

For a line with a given integrated atomic absorption cross section, s, the amount of absorption or emission at a particular frequency will depend on the width of the line over which s is spread. We introduce an absorption profile function, $\phi(\Delta\nu)$, such that

$$s_\nu = s\phi(\Delta\nu) \tag{2-29}$$

where $\Delta\nu$ equals $\nu - \nu_{jk}$, the distance from the line center in frequency units. From equation (2-21) it is evident that the integral of $\phi(\Delta\nu)\,d\nu$ over a line equals unity. From equations (2-2), (2-20), and (2-29) we see that τ_ν along the line of sight is related to $\phi(\Delta\nu)$ by the expression

$$\tau_\nu = N_j s\phi(\Delta\nu) \tag{2-30}$$

where N_j is the integral of the number of absorbing atoms (those in the lower state j) along the line of sight; i.e., N_j is the number of such atoms in a cylindrical column 1 cm^2 in cross section, extending the full length of the line of sight.

A situation of basic importance is that in which the line profile is Maxwellian, and

$$\phi(\Delta\nu) = \frac{1}{\beta\pi^{1/2}} e^{-(\Delta\nu - \Delta\nu_0)^2/\beta^2} \tag{2-31}$$

where $\Delta\nu_0$ is the shift of the line center, resulting from the mean relative velocity of the absorbing atoms with respect to the observer. In terms of the dispersion of w_r, the radial velocity of the atoms (or molecules), the quantity β is given by

$$\frac{\beta^2}{v^2} = \frac{2(\langle w_r^2 \rangle - \langle w_r \rangle^2)}{c^2} \tag{2-32}$$

where the brackets, $\langle \ \rangle$, denote a mean value over a Maxwellian distribution. If the dispersion of atomic velocities is due to random thermal motions,

$$\frac{\beta^2}{v^2} = \frac{2kT}{m_a c^2} \tag{2-33}$$

where m_a is the atomic mass.

The detailed measurement of line profiles is frequently difficult, and for absorption lines it is observationally simpler to measure the equivalent width, defined in frequency units as

$$W = \int \left(1 - \frac{I_\nu}{I_{\nu c}}\right) d\nu \tag{2-34}$$

where $I_{\nu c}$ is the intensity in the adjacent continuous spectrum, and I_ν is the intensity in the line; the integral again extends over the line in question. For measures of integrated starlight the flux $\mathscr{F}_\nu$ replaces the intensity I_ν in this expression. If, in accordance with observational usage, W is defined in wavelength units, the value obtained from equation (2-34) and subsequent equations must be multiplied by λ/ν.

Since the radiation being absorbed is usually concentrated in a very small solid angle, the emission, j_ν, which is generally isotropic, is negligible in equation (2-3); if $I_{\nu c}$ is set equal to $I_\nu(0)$, equation (2-34) yields

$$W = \int (1 - e^{-\tau_\nu}) d\nu \tag{2-35}$$

If τ_ν is small we may expand the exponent, retaining only the first two terms and find

$$W = \int \tau_\nu d\nu = N_j s \tag{2-36}$$

2. OBSERVATIONS OF THE GAS

where we have used equation (2-30). If we denote by W_A the corresponding equivalent width in Ångstrom units, and neglect stimulated emission, equations (2-36) and (2-25) yield

$$W_A = 8.85 \times 10^{-13} \lambda_\mu^2 f_{jk} N_j \qquad (2\text{-}37)$$

where λ_μ is the wavelength in microns. The general relation between W and N_j is called a "curve of growth" and equation (2-37) gives the so-called "linear section" of this curve, valid when across the entire line profile τ_v is small compared to unity.

For τ_v large, I_v/I_{vc} becomes very small and the amount of energy absorbed is no longer proportional to τ_v; the line is said to be "saturated" at least at its center. In this case the curve of growth depends on the detailed shape of $\phi(\Delta v)$. If equation (2-31) is substituted into equation (2-35), straightforward algebra leads to the result

$$W = 2\beta F(C) \qquad (2\text{-}38)$$

where

$$F(C) \equiv \int (1 - e^{-C \exp(-x^2)}) \, dx \qquad (2\text{-}39)$$

with C, the value of τ_v at the line center, equal to

$$C = \frac{N_j s}{\pi^{1/2} \beta} \qquad (2\text{-}40)$$

If equations (2-38) and (2-40) are combined, we obtain in place of equation (2-37) the more general result

$$W_A = \left\{\frac{2F(C)}{\pi^{1/2} C}\right\} 8.85 \times 10^{-13} \lambda_u^2 f_{jk} N_j \qquad (2\text{-}41)$$

For small C a series expansion of $F(C)$ gives equation (2-37); for C large equation (2-39) gives the asymptotic result

$$F(C) = (\ln C)^{1/2} \qquad (2\text{-}42)$$

When equation (2-42) is applicable, an equivalent width is said to be on the "flat section" of the curve of growth. Numerical values of $F(C)$, computed from series expansions [4], are given in Table 2.1. Also tabulated in Table 2.1 are the value of $F(2C)/F(C)$, the

TABLE 2.1

Curve of Growth for Doppler-Broadened Absorption Lines

C	0.000	0.10	0.20	0.30	0.40	0.50	0.60	0.80
$F(C)$	0.000	0.086	0.165	0.240	0.309	0.374	0.435	0.545
$F(C)/C$	0.886	0.856	0.827	0.799	0.774	0.749	0.725	0.682
$F(2C)/F(C)$	2.000	1.932	1.871	1.815	1.763	1.716	1.674	1.598
C	1.0	1.2	1.4	1.6	2.0	3.0	4.0	6.0
$F(C)$	0.643	0.728	0.804	0.872	0.986	1.188	1.320	1.483
$F(C)/C$	0.643	0.607	0.575	0.545	0.493	0.396	0.330	0.247
$F(2C)/F(C)$	1.535	1.481	1.436	1.398	1.338	1.248	1.200	1.156
C	10	20	30	40	60	100	1000	10,000
$F(C)$	1.66	1.86	1.97	2.04	2.14	2.26	2.73	3.12
$F(C)/C$	0.166	0.093	0.066	0.051	0.036	0.0226	0.0027	0.0003
$F(2C)/F(C)$	1.123	1.098	1.087	1.081	1.074	1.067	1.046	1.035

"doublet ratio" for two lines for which τ_v is in the ratio of 2 to 1. As C increases both lines become fully saturated, the doublet ratio approaches unity, and $F(C)/C$ decreases.

For sufficiently strong lines the radiation damping wings should become important. In the limiting case that these wings dominate we may write

$$\phi(\Delta v) = \frac{\gamma/\pi}{\gamma^2 + (\Delta v)^2} \tag{2-43}$$

where γ, a measure of the width of the upper level, is determined from the Einstein coefficients through the relation

$$\gamma = \frac{1}{4\pi} \sum_j A_{kj} \tag{2-44}$$

summed over all lower levels. Substitution of equations (2-30) and (2-43) into (2-35) and direct integration yield

$$W = 2(Ns\gamma)^{1/2} \tag{2-45}$$

2. OBSERVATIONS OF THE GAS

In deriving this result the term γ^2 in the denominator of equation (2-43) has been ignored, an approximation valid for strong lines.

Equation (2-45) is applicable to lines on the "square-root section" of the curve of growth. Values of W have been computed also [5] for intermediate cases, when both radiation damping and Doppler broadening must be considered.

Continuous Emission and Absorption by Hydrogenic Atoms

For electrons with a Maxellian velocity distribution relative to the protons the emission coefficient for free-free transitions (bremsstrahlung) is given by the usual formula [6]

$$j_v = \frac{8}{3}\left(\frac{2\pi}{3}\right)^{1/2}\frac{e^6}{m^2c^3}\left(\frac{m}{kT}\right)^{1/2} g_{ff} n_e n_p e^{-hv/kT}$$
$$= 5.44 \times 10^{-39}\frac{g_{ff} n_e n_p}{T^{1/2}} e^{-hv/kT} \text{ erg cm}^{-3} \text{ sec}^{-1} \text{ sr}^{-1} \text{ Hz}^{-1}$$
(2-46)

Here g_{ff}, which varies slowly with frequency, v, is the Gaunt factor for free-free transitions [7]. The corresponding emissivity for free-bound transitions to a level of principal quantum number n is smaller by approximately $-4E_n/kT$, when kT much exceeds the binding energy, $-E_n$, of level n. Since $-E_n$ must be less than hv, free-bound emission is negligible for radio frequencies, for which hv/kT is very small.

The total amount of energy radiated in free-free transitions per cm^3 per sec, which we denote by ε_{ff}, is obtained on multiplying equation (2-46) by 4π and integrating over dv. We obtain

$$\varepsilon_{ff} = \left(\frac{2\pi kT}{3m}\right)^{1/2}\frac{2^5\pi e^6}{3hmc^3} n_e n_p \bar{g}_{ff}$$
$$= 1.42 \times 10^{-27} n_e n_p T^{1/2} \bar{g}_{ff} \text{ ergs cm}^{-3} \text{ sec}^{-1} \quad (2\text{-}47)$$

Corresponding to the emissivity in free-free transitions there will also be an absorption coefficient, κ_v, related to j_v by Kirchhoff's law, equation (2-5). To obtain κ_v for radio waves we substitute equations (2-46) and (2-4) into equation (2-5), insert the value [7] of g_{ff} for v small compared to kT/h but large compared to the plasma frequency, and obtain

$$\kappa_v = \frac{4(2\pi)^{1/2}e^6}{(3mk)^{3/2}c}\left[\frac{3^{1/2}}{\pi}\left\{\ln\frac{(2kT)^{3/2}}{\pi e^2 v m^{1/2}} - \frac{5\gamma}{2}\right\}\right]\frac{n_e n_p}{T^{3/2}v^2}$$

$$= 0.173\left\{1 + 0.130\log\frac{T^{3/2}}{v}\right\}\frac{n_e n_p}{T^{3/2}v^2}\ \mathrm{cm}^{-1} \qquad (2\text{-}48)$$

where γ is Euler's constant, equal to 0.577, and log denotes logarithms to base 10; the square brackets enclose g_{ff}.

Synchrotron Emission

An electron gyrating in a magnetic field emits radiation. When viewed from a frame of reference where the electron is instantaneously at rest, the energy radiated has a wide angular distribution, with a maximum perpendicular to the direction of acceleration. When transformed to a reference frame where the electron has a large relativistic energy, E, much exceeding the rest mass energy mc^2, most of the energy appears in photons emitted in very nearly the same direction as the instantaneous velocity of the particle. As the relativistic electron gyrates in a circle, this cone of radiation, which will sweep by an observer very rapidly, will contain very high frequencies. The frequency of maximum intensity is about a third of v_c [8], where

$$v_c = \frac{3eB_\perp}{4\pi mc}\left(\frac{E}{mc^2}\right)^2 \qquad (2\text{-}49)$$

$B_\perp$ is the component of the magnetic field, **B**, perpendicular to the particle velocity, and m is the electron rest mass. If $P(E, v)\,dv$ represents the total energy radiated per second within the range of frequency dv by an electron of energy E,

$$P(E, v) = \frac{3^{1/2}e^3 B_\perp}{mc^2}F(v/v_c) \qquad (2\text{-}50)$$

where $F(x)$ is a function which varies as $x^{1/3}$ for small x, and as $x^{1/2}e^{-x}$ for large x; its maximum value of 0.92 is reached at x equal to 0.29 [8].

2. OBSERVATIONS OF THE GAS

If the particle density $n_e(E) \, dE$ for energetic electrons within an energy range dE is given by a power-law distribution, as is observed for cosmic-ray particles, we may write

$$n_e(E) = KE^{-\gamma} \tag{2-51}$$

A straightforward integration of $n_e(E)P(E, v) \, dE$ over all dE gives the total rate of emission of energy per second per cm^3 per frequency interval. We divide this quantity by 4π to obtain j_v, on the assumption that the radiation field is isotropic; such a uniform distribution of emitted intensity over all solid angles would result if the magnetic field **B** were randomly oriented. If also j_v is assumed uniform over a distance L, then an integration of equation (2-1) over distance, with neglect of absorption, gives I_v produced by synchrotron emission. The corresponding brightness temperature, computed from equation (2-7), is given by

$$T_b = \frac{Le^3 B}{2mk} \left(\frac{3eB}{4\pi m^3 c^5}\right)^{(\gamma-1)/2} \frac{Ka(\gamma)}{v^{2+(\gamma-1)/2}} \tag{2-52}$$

In deriving equation (2-52) an isotropic distribution of particle velocities at each point has been assumed. The function $a(\gamma)$ varies [8] from 0.103 to 0.074 as γ increases from 2 to 3, the range of relevant values. Since $(\gamma - 1)/2$ is generally not large, synchrotron radiation produces a specific intensity which varies slowly with frequency while the corresponding brightness temperature decreases rapidly with increasing v. Synchrotron radiation tends to be polarized with the electric vector perpendicular to **B**. If the velocity distribution is isotropic and the distribution of energies follows equation (2-51), the emitted radiation is about 70% polarized for γ between 2 and 3, and for **B** uniform [8]. If the magnetic field is randomly oriented along the line of sight the polarization disappears.

The rate at which a relativistic electron loses energy by synchrotron radiation may be computed by integrating $P(E, v) \, dv$ in equation (2-50) over all dv. Since the integral of $F(x) \, dx$ from 0 to ∞ equals $8\pi/9(3)^{1/2}$, we obtain

$$\frac{E}{-dE/dt} = \frac{3mc^3}{2e^2}\left(\frac{mc}{eB_\perp}\right)^2 \frac{mc^2}{E} = \frac{16.3}{B_\perp^2}\frac{mc^2}{E} \text{ years} \quad (2\text{-}53)$$

2.2 Emission from the Gas

A number of major features in the interstellar emission spectrum are due to hydrogen, the chief constituent of the gas. The 21-cm line of neutral H is emitted primarily from H I regions (see Section 1.1). In H II regions collisions between electrons and protons lead to emission of continuous radiation, as well as to recombination; the neutral H atoms formed are frequently in excited levels, and radiate both in visual and radio regions of the spectrum. Emission lines from other atoms and ions are also produced in H II regions, while molecular OH lines seem to be emitted in large part from boundary layers between H I and H II regions. We summarize very briefly here the chief results obtained in these active observational fields.

Hydrogen 21-cm Line

For a hydrogen atom in the ground electronic level the total energy is some 5.9×10^{-6} eV higher if the electron and proton spins are parallel ($g_k = 3$) than if they are antiparallel ($g_j = 1$). A radiative transition between these two adjacent states produces radiation at 1420.406 MHz, with a corresponding wavelength of about 21.11 cm. The computed spontaneous transition probability [9] for this magnetic dipole radiation is

$$A_{kj} = 2.868 \times 10^{-15} \text{ sec}^{-1} = \frac{1}{1.10 \times 10^7 \text{ years}} \quad (2\text{-}54)$$

The corresponding upwards oscillator strength, computed from A_{kj} by combining equations (2-22), (2-24), and (2-25) is 5.75×10^{-12}. If $n_k/n(\text{H I})$, the fraction of the total number of neutral H atoms, per cm^3, that are in the upper state of the transition, is set equal to 3/4, equal to the ratio $g_k/(g_j + g_k)$ of the statistical weights, the integral of the emissivity across the line is obtained directly from equation (2-16), and is evidently independent of

2. OBSERVATIONS OF THE GAS

temperature. Since self-absorption is important, however, we must compute the optical depth from equation (2-30) and then determine I_v from equation (2-6). The line broadening is entirely due to the velocity distribution of the H atoms; hence we evaluate $\phi(\Delta v)$ in equation (2-30) in terms of $P(v) \, dv$, the fraction of atoms whose radial velocities lie within the range dv. Since $P(v) \, dv$ equals $\phi(\Delta v) \, dv$, and dv/dv equals c/v, we obtain

$$\phi(\Delta v) = \frac{c}{v} P(v) \qquad (2\text{-}55)$$

The integrated atomic absorption cross section, s, is determined from equation (2-28), with the usual assumption that b_k/b_j is exactly unity. The frequent collisions which tend to establish a Maxwellian distribution at the kinetic temperature, and thus yield $b_j = b_k$, will be discussed in Section 4.2; however, other mechanisms could conceivably alter the ratio of b_k to b_j. Further information, either theoretical or experimental, on the value of b_k/b_j would be desirable.

If these results are combined, we obtain for τ_v

$$\tau_v = 5.49 \times 10^{-14} \frac{N(\text{H I}) P(v)}{T} \qquad (2\text{-}56)$$

where $N(\text{H I})$ denotes the number of neutral H atoms per cm^2 in the line of sight. In equation (2-8) we shall ignore T_{b0}, the brightness temperature of the radiation incident on the hydrogen, and regard T_b as a function of the velocity shift v, related to Δv by the usual Doppler shift formula. If τ_v is small, the exponent in equation (2-8) can be expanded, retaining again only the first two terms. If $T_b(v)$ is integrated over the velocity dv (in cm/sec) we obtain

$$N(\text{H I}) = 1.82 \times 10^{13} \int T_b(v) \, dv \text{ cm}^{-2} \qquad (2\text{-}57)$$

Observed values [10] of $T_b(v)$ in the galactic plane are shown in Figure 2.1. The double sine-wave pattern shown by the observations results from the radial velocity due to differential galactic

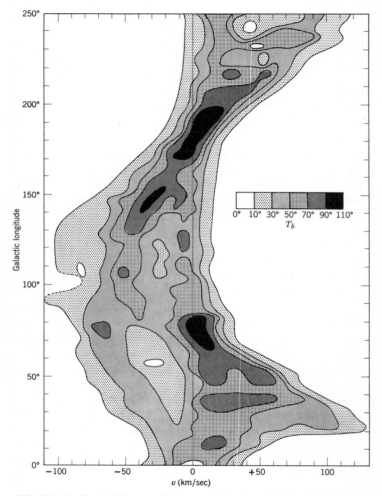

Fig. 2.1. Surface brightness of 21-cm emission in galactic plane. The contours, taken from ref. [10], give values of the brightness temperature, T_b, of 21-cm radiation in the galactic plane as a function of the radial velocity, v, (corrected for solar motion) and of galactic longitude, l.

2. OBSERVATIONS OF THE GAS

rotation. Except at the nulls of differential rotation the magnitude of the shift can be used to determine roughly the position of the emitting gas in the Galaxy. In this way the pattern of spiral arms within the Galaxy has been delineated [10, 11] and densities of neutral H determined [12]. Within the Orion arm, which passes by the Sun, $n(\text{H I})$ rises to between 1 and 2 cm^{-3}, while in the adjacent Perseus arm about 2 kpc away a density between 2 and 3 cm^{-3} seems indicated. If self-absorption were greater than assumed, $n(\text{H I})$ would be increased. On the other hand, if b_k/b_j differs appreciably from unity, the computed $n(\text{H I})$ could be in error in either direction. If an average is taken in the galactic plane, including regions between spiral arms as well as the arms themselves, the average density of neutral H at distances between 5 and 10 kpc from the center appears to be [11] about 0.5 cm^{-3}. The measured brightness temperature for radiation perpendicular to the plane, combined with this mean density in the plane, gives an effective thickness, 2H, of the hydrogen gas equal to about 220 pc [11]; this effective thickness is defined as $N(\text{H I})$, the total number of neutral hydrogen atoms in a cylinder 1 cm^2 in cross section extending through the Galaxy, divided by the neutral hydrogen density $n(\text{H I})$ in the plane of symmetry of the Galaxy.

Only rather preliminary results are available on the fine-scale distribution of hydrogen density in the general neighborhood of the Sun. Because of random velocities of fluid elements in the gas, the observed radial velocity cannot be attributed entirely to differential galactic rotation, and hence cannot be used to give a precise location for the emitting atoms. As a result, the density of neutral H between the spiral arms is not well known, although the low values of T_b for some intermediate values of v suggest that the inter-arm density may be less than 0.1 cm^{-3}. Changes in the observed $T_b(v)$ as either the velocity or the direction is changed show that the interstellar hydrogen gas is not very uniform. However, a detailed study at intermediate latitudes [13] indicates that most of the emitting atoms in the region studied are in a rather sharply bounded but otherwise relatively uniform structure several hundred parsecs long, with only a few percent of the mass in clouds and cloudlets a few parsecs in size. The mean particle density of H atoms in this region is in the range 0.1 to 0.3 cm^{-3} at a height of several hundred parsecs above the galactic plane. The kinetic temperature is difficult to determine from the 21-cm data. The maximum brightness temperature observed [10] is about 125°, found in the direction of the galactic center and along the Orion arm. According to equation (2-8), based on a uniform value of the kinetic temperature, T, the maximum brightness temperature, T_b, is just equal to T, if T_{b0} can be neglected. On this picture a kinetic temperature of 125° would be consistent with these observations. If the temperature varies with position in H I regions, a different interpretation of this maximum intensity is required. If the scale of the variability is very small, so that the line of sight of each element samples about the same mixture of hot and cold elements, then the harmonic mean temperature must be used [14] in equations (2-56) and (2-8). On the other hand, if T is assumed to be constant along each line of sight but different in different directions, and τ_v is large in each direction, the observed T_b will be a linear average of T over all directions included in the beam. The real case is likely to be more complex than either model.

Attempts to correlate 21-cm emission with the presence of obscuring dust have yielded information which suggests that T may differ between different regions. In a general way, the dust and H atoms seem to occur together, as shown by the correlation between T_b and low counts of external galaxies [15]. However, more detailed studies, especially of denser clouds, can show a negative correlation. For example in the direction of an obscuring cloud in Auriga, which star counts indicate as being roughly 2 pc × 8 pc in size, at a distance of some 200 pc, T_b at 21 cm is only about 60° as compared to about 70° in most of the surrounding region [16]. Since most of the emitting H atoms should be much further away than 200 pc, a cold absorbing cloud, with T less than 60°, seems indicated.

The measures indicate an appreciable dispersion of velocities within the H gas. Measures of $T_b(v)$ toward the center or anticenter indicate an "external" velocity dispersion σ_e, equal to

2. OBSERVATIONS OF THE GAS

$\langle v^2 \rangle_e^{1/2}$, of about 6 km/sec, with values up to 9 km/sec obtained from measures in other directions at a velocity shift opposite to that predicted from galactic rotation [17]. At high latitudes, where individual components of an emission line can be resolved, without too much blending, it is possible to measure the internal dispersion (σ_i or $\langle v^2 \rangle_i^{1/2}$) of atomic velocities within a single cloud or structure in the turbulent gas. The most likely value of σ_i is between 2 and 3 km/sec, with only a few percent of the clouds yielding values of σ_i in the range from 1 to 1.5 km/sec [18]. A comparison of these values of σ_i with those obtained from the 21-cm line in absorption is given in the following section.

The presence of some high-velocity clouds at high galactic latitudes is an unexpected result [19], with N(H I) rising as high as 10^{20}/cm^2 for velocities between -30 and -80 km/sec, with about one-fifth this number at the corresponding positive velocities (relative to the local standard of rest). The few even faster clouds are approaching almost without exception. These high velocity clouds tend to be concentrated at galactic longitudes between about 100° and 140°. On the average, some 10^{19} H atoms/cm^2 are observed approaching the galactic plane at a mean velocity of about 75 km/sec.

Measures of emission-line profiles in the neighborhood of stellar associations show shells of expanding hydrogen around some of these. In particular, around the Orion nebula [20] and about λ^1Ori [21] are shells of gas expanding at 10 and 8 km/sec, respectively, with masses of about 5×10^4 $M_\odot$. The former shell is more extended and less dense, with inner and outer diameters of 20 and 80 pc, as compared to 15 and 35 pc for the latter. On the other hand, an investigation of 21-cm emission for 27 H II regions revealed [22] appreciable neutral hydrogen only around 4 of these, and with line widths so narrow that any expansion velocity could not exceed about 3 km/sec.

Visual Emission Lines

Faint emission of the Balmer lines from extended regions around early-type stars has been observed for some time [23, 24]. In addition to $H\alpha$, $H\beta$, and $H\gamma$ faint forbidden lines of O III at 5006.8 and 4958.9 Å (designated N_1 and N_2 because of their original attribution to an unidentified "nebulium") and of the corresponding lines from the isoelectronic ion, N II, at 6548.1 and 6583.6 Å are generally observed, as are the forbidden lines of O II at 3726.1 and 3728.8 Å. Emission is almost always observed near O stars within 250 pc of the galactic plane [25].

According to equation (2-18) the emissivity in the Balmer lines varies as $b_k n_e^2$, if the proton and electron densities are assumed equal. If absorption is negligible, but j_v is uniform over a region of length L, then I_v equals $j_v L$, as is evident from equation (2-1). Hence I_v varies as $b_k n_e^2 L$, multiplied by a known function of temperature, T. The "emission measure," E_m, is defined as

$$E_m = n_e^2 L \text{ pc/cm}^6 \qquad (2\text{-}58)$$

where n_e is measured per cm^3, with L in parsecs. In interpreting the measured values of I_v for the Balmer lines it is customary to assume a particular temperature, usually 10,000°, to take a theoretical value for b_k at that temperature, and then to compute E_m from I_v with use of equations (2-18) and (2-13). A correction is often required for interstellar extinction, which may be rather uncertain. The resultant values of E_m range [24] from about 200 for the faintest regions on which photographic photometry is possible up to nearly 10^7 for the line of sight through the center of the Orion nebula, the brightest diffuse nebula. If L is estimated and n_e is assumed uniform, the resultant values of n_e range from about 5 cm^{-3} in the faintest objects up to about 10^3 cm^{-3} in the Orion nebula. Of the 33 values of n_e obtained in this way about half are less than 30 cm^{-3}; the true median value is presumably less because of observational selection.

The value of b_k used in computing the emission measure is based on the equations of statistical equilibrium discussed in Section 4.2. For example, a detailed application of these equations [26] to the case of hydrogen atoms formed by radiative capture, taking into account the variation of b_k between states of different angular momentum, gives 0.13 for the mean b_4 effective in the

2. OBSERVATIONS OF THE GAS

production of $H\beta$, when the electron temperature is 10,000°; this value is based on the assumption that any quanta emitted in the Lyman lines are promptly reabsorbed by the neutral atoms. While the theory does not seem to agree very well with the observed Balmer decrement [26], the values of n_e^2 obtained with the computed b_k are probably not greatly in error. A comparison of these results with those obtained by radio continuum observations is given below.

Emission lines from elements other than H have been somewhat less extensively observed because of their relative weakness. Ratios of intensities for different lines produced by the same ion may be used for direct determinations both of n_e and of T. For example, the ratio of intensities of the two O II lines at $\lambda 3729$ and $\lambda 3726$, produced by emission from adjacent levels with an energy separation of only about 0.002 eV, depends primarily on the electron density and very little on the electron temperature. According to the theory presented in Section 4.2 [equation (4-39)], at high n_e collisional deexcitations tend to balance collisional excitations and the ratio of population densities approaches that in thermodynamic equilibrium; from equation (2-18) we see that the total intensity in the line varies as $g_k A_{kj}$. At low n_e, on the other hand, the number of photons emitted is limited to the number of collisional excitations, and the line intensities vary as the excitation cross sections, σ_{jk}. Thus in the Orion nebula n_e was found [27] to decrease from about 1.5×10^4 cm^{-3} near the center of the Trapezium to 2.5×10^2 cm^{-3} at a distance of some 25'. This is some six times higher than the electron density obtained from the emission measure. These two results can be reconciled if the gas is clumpy, with regions of high density surrounded by regions of low density. To adopt a simplified theoretical model, let all the gas be concentrated in clumps or clouds occupying a fraction, F, of the total volume of the nebula. If n_{ec} is the electron density in the clouds, and the density between the clouds is negligible, then

$$\langle n_e^2 \rangle^{1/2} = F^{1/2} n_{ec} = \frac{\langle n_e \rangle}{F^{1/2}} \qquad (2\text{-}59)$$

where the brackets denote mean values over the volume of the nebula. Thus a value of 1/36 for F in the Orion nebula would be consistent with a measured n_{ec} equal to six times the r.m.s. electron density found from the emision measure. Since the ratios of line intensities become insensitive to n_e, for values of n_e less than 10^2 cm^3, it is not known whether similar violent density fluctuations are present in other, less dense H II regions.

The ratio of the two O III emission lines at $\lambda 4363$ and $\lambda 5007$ (N_1), whose upper levels differ in excitation energy by 2.8 eV, varies primarily with the Boltzmann factor in equation (2-9), multiplied by the ratio of the b_j factors involved; this latter ratio may be computed from equation (4-39) and T then found from the observed intensity ratio of the two lines. Values of T determined [28] in this way are 9000°K for the Orion nebula and 7500° to 8200° for two H II regions in which the electron density computed from E_m (with no allowance for possible density fluctuations) is about 10^2. Electron temperatures in the range from 6000° to 9000° have been determined [23, 29] from the ratio of O II, N II, and O III lines to $H\alpha$, but this result depends on assumptions concerning the relative abundances of the different emitting atoms. However, for the cooler nebulae the oxygen tends to be singly ionized and the line ratios of the O III lines cannot be observed.

Velocities of the emitting gases have been measured spectroscopically in a few cases where line intensities are sufficiently great to permit high resolution. In M16 the hydrogen gas in the bright rims has been observed [30] to be approaching at about 13 km/sec relative to the gas at neighboring points. Detailed measures in the Orion nebula [31] indicate a general expansion of the gas at about 10 km/sec, with line doubling in some regions, suggesting shocks with relative velocities as great as 25 km/sec. The widths of the emission lines do not vary with the inverse square root of the atomic weight, as would be expected if thermal motions were involved; if a kinetic temperature of 10,000°K is assumed the r.m.s. turbulent velocity required along the line of sight has a minimum value of about 6 km/sec for O^{++} and 8 km/sec

2. OBSERVATIONS OF THE GAS

for H, somewhat less than the sound velocity of 11 km/sec (see Table 5.1).

Velocities of gaseous shells produced by supernova explosions have been determined [32] from Doppler shifts of emission lines in the visual as well as from the observed size together with the time since the initial explosion. In the Crab nebula, produced by the supernova of 1054 A.D., the Doppler shifts indicate an expansion velocity of about 1000 km/sec, which is apparently about equal to the average velocity since the explosion. For the condensations observed in the radio source Cas A the observed velocities of expansion are about six times greater. For the remnants of Tycho's supernova in 1572 A.D.—believed to be a typical type I supernova—the average velocity deduced from the angular size, the distance, and the expansion time is about 13,000 km/sec. On the other hand, in the Cygnus Loop, which is apparently only one section of an expanding shell produced by a much older supernova, the velocities are much less, in the neighborhood of 100 km/sec [32]. In these various remnants the mass of the expanding gas, as computed from the emission measure, is generally in the range from 0.1 to 1 $M_\odot$.

Radio Continuum and Line Emission

The continuous radio emission from H II regions has been observed at various radio frequencies, with extensive measures [33] at 1390 MHz (21.6 cm). For the Orion nebula and two other relatively bright objects the radio spectra have been observed over a factor of 50 to 100 in wavelength, and are in general agreement [34] with theoretical expectations for free–free transitions; the intensity is independent of frequency for the shorter wavelengths, for which τ_ν is small, but varies as ν^2 in accordance with equations (2-7) and (2-8) for the longer wavelengths, characterized by a larger optical depth. Measures of radio emission from regions of ionized hydrogen should, in principle, give more accurate values of E_m than do the optical measures, since no correction is necessary for extinction by solid particles or for deviations from thermodynamic equilibrium; the theoretical evaluation of the Gaunt factor, g_{ff}, which appears in equation (2-46) should be reliable. A comparison [24] of the values of n_e obtained from the radio and visual measures shows that the mean ratio of these values, averaged over 19 H II regions, equals 1.4, with an average deviation of 0.5, and with individual values of n_e ranging from 5 to 700 cm^{-3}. Evidently the agreement between these two independent methods is relatively good, considering the uncertainties. Systematic differences between the average of n_e^2 and of n_e caused by clumpiness in the gas [see equation (2-59)] would affect both optical and radio observations equally, and the mean electron density may be systematically less than the values found, possibly by the same factor 6 found in the Orion nebula [27]. Any difference between the actual temperature and the 10,000° assumed in computing E_m from the measured fluxes [see equations (2-18) and (2-46)] would change n_e somewhat differently for the two sets of data.

The overall distribution of ionized hydrogen in the Galaxy has been determined [33] from the general intensity of the continuous emission at 1390 MHz. The separation of the thermal and nonthermal components, discussed in Section 2.4, gives rise to difficulties, and the corrections for nonthermal sources may not be unique. Between 7 and 10 kpc from the center of the Galaxy the r.m.s. proton density is found to be 0.45 cm^{-3}, rising to about 0.8 cm^{-3} in the Orion arm near the Sun. As shown in Section 2.4 this result is comparable with the values obtained from the measured absorption of nonthermal radiation by the hydrogen gas. The mean electron density is less than the r.m.s. value by $F_i^{1/2}$, where F_i is the fraction of space occupied by the denser H II regions of ionized hydrogen which make the dominant contributions to the emission and absorption; estimates of F_i have ranged from 10^{-3} up to 5×10^{-2}. Since there is no reliable evidence on the value of F_i, the observed intensity of the thermal radio emission does not give the mean value of n(H II), the density of ionized hydrogen, but does give useful information on the total ultraviolet luminosity of early-type stars (see Section 5.1).

2. OBSERVATIONS OF THE GAS

The strongest H recombination lines expected in the radio regime [35] are the "α transitions" from n to $n - 1$; i.e., between levels whose total quantum number differs by unity. The value of $A_{n,n-1}$ for such transitions is $5.2 \times 10^9\ n^{-5}\ \text{sec}^{-1}$; which for n about equal to 10^2 is very much greater than for the 21-cm line. However, the number density of atoms in these highly excited states is low, and the resultant brightness temperature, measured for lines with n equal to 91, 105, 109, 110, 157, and 159, is less than a degree above background. A comparison of equations (2-18) and (2-46) shows that the ratio of the integrated emissivity over such a line to the emissivity in the adjacent continuum should vary as $b_n T^{1/2}/f_e$ or, from equation (2-13), as b_n/T, with a known constant of proportionality. As we have seen from equation (2-28) if b_n differs somewhat from b_{n-1} for a line with small $h\nu/kT$, the effect of self-absorption is much altered and may be either positive or negative, complicating the interpretation of the observations.

Measurements of the 109α line and of the adjacent continuum in some dozen H II regions [36] have been used to give kinetic temperatures on the assumption that b_{109} and b_{108} are both equal to unity. This assumption seems consistent with the observational evidence, but is open to some question theoretically [37]. Individual values of T found in this way lie in the range from about 4000° to 10,000°, with an average value of 5800°. The average internal velocity dispersion of 19 km/sec obtained from the line widths is substantially greater than the random thermal velocities, as would be expected for expanding clouds of gas.

OH Emission

The four OH radio lines at 1612, 1665, 1667, and 1721 MHz [12,38] have spontaneous transition probabilities in the range from 10^{-11} to $10^{-10}\ \text{sec}^{-1}$; these lines have been observed in various directions in the Galaxy, sometimes in absorption and sometimes in emission. Their behavior appears to be quite different from that of the other lines observed in interstellar space. A detailed survey of some fifty H II regions in the southern Milky Way shows [38] OH emission in about half these. However, the emission comes only from small zones, with an angular diameter less than 0.1 arcsec, yielding diameters less than a small fraction of a parsec and true brightness temperatures greater than $10^{9\circ}$K; in contrast, the width of the emission peak is very narrow, corresponding to Doppler broadening at a temperature less than 50°K, and in some cases as low as 4°K. The relative intensities of the four lines are wildly different from the theoretical values of $g_k A_{kj}$, and differ from one zone to another. In some sources virtually 100% circular polarization is observed; in others, up to 40% plane polarization. In general these emission zones tend to be located at the apparent edge of H II regions.

All these observations are consistent with the assumption that the emission lines are produced by maser action, corresponding to negative absorption produced by a population inversion, and a resultant b_k/b_j greater than unity in equation (2-28). If the circular polarization is attributed to a magnetic field, B must equal about 10^{-3} G, a value which for the interstellar medium generally seems too high to reconcile with any existing theories. The observation that in one case the relative intensities changed during one week would, if confirmed, give an important clue to the origin of the population inversion.

2.3 Absorption by the Gas

The information obtained from absorption lines complements in an important way that obtained from the emission lines. In the visual spectrum the lines seen in absorption are produced by entirely different atoms from those seen in emission; because of the small number of atoms required to produce a measurable absorption line, these lines provide a sensitive tool for measuring the gas velocities in different directions. In the radio spectrum the 21-cm line is seen both in absorption and emission. Comparison of these two types of data gives evidence for variations within the gas. The results obtained in these two regions of the spectrum are summarized below.

2. OBSERVATIONS OF THE GAS

Radio Lines

The 21-cm line has been seen in absorption in the spectra of various radio sources [12]. Equation (2-8) may be used to compute τ_v, with T_{b0} determined from intensity measures of the source at adjacent wavelengths, and the emission contribution $T(1 - e^{-\tau_v})$ determined from frequency scans of the emission profile in directions adjacent to the source. From equation (2-56) it is evident that τ_v varies as $1/T$, and hence if N(H I) is known from the emission components, T can be found from τ_v measured in absorption.

However, the apparent differences between the absorption and emission profiles suggest that these two sets of lines may be formed in different regions [39, 40]. For example, the line of sight to the Cas A source passes through about 1 kpc of the Orion arm. The 21-cm emission line in directions adjacent to Cas A has a brightness temperature corresponding to 2×10^{21} atoms/cm^2 and a total velocity width of some 15 km/sec. The 21-cm absorption line profile in the Cas A radio source shows weak absorption over this width, but a temperature of about 500° is required to fit the value of N(H I) observed from the emission. On the other hand, relatively much stronger absorption is seen in Cas A in a relatively narrow line at nearly zero velocity in the local standard of rest. A temperature of 50° in the absorbing gas would account for strong absorption, even though this cloud is not strongly seen in emission.

Another indication of possible differences between emitting and absorbing clouds of hydrogen is provided by comparison of the values of σ_i, the internal velocity dispersion for the components of emission and absorption lines. Such a comparison is given in Figure 2.2, where the distribution of σ_i is plotted for emission line components observed at values of b, the galactic latitude, between 10° and 20° [18] and for absorption lines in 15 radio sources [40], all but three of which have $|b|$ less than 10°. The band widths used for the emission and absorption observations were about the same (5 and 6 kHz, respectively) and the methods used for analyzing the profiles in terms of Gaussian components were similar. For

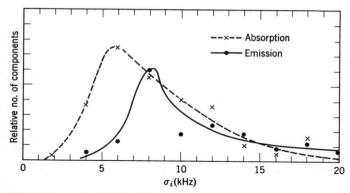

Fig. 2.2 Internal velocity dispersion in 21-cm emission and absorption lines. The two curves show relative numbers of components with different widths. The horizontal scale is in units of 10^3 Hz; the two curves are normalized to the same area, including components with σ_i greater than 20 kHz. The data are taken from refs. [18] and [40].

the emission line components about 34% lie off the figure, at σ_i greater than 21 kHz, as compared with only 7% for the absorption components. The large difference in distribution is obvious, though some of this may be caused by the different distances from the galactic plane. Emission components cannot be measured precisely at low galactic latitudes because of overlapping.

The observations are generally consistent with the picture that two types of hydrogen clouds are present—relatively rapidly moving clouds at a higher temperature and slowly moving clouds at a lower temperature, with comparable amounts of mass in the two types. In the actual situation a continuous distribution of types is more likely. Much more detailed information is needed to indicate how variations in the density, temperature, and velocity of the interstellar gas are correlated with each other and with location in the Galaxy. The temperature deduced from the 21-cm observations may differ from the kinetic temperature if the ratio of the b_j values for the hyperfine levels in H differs from unity (see Section 4.2).

The intensities and widths of the OH lines seen in absorption have been used [12] to obtain information on the temperature and

2. OBSERVATIONS OF THE GAS

composition of the gas. However, the observations discussed in the previous section indicate that maser amplification influences the emission lines; as a result, the interpretation of the absorption line intensities seems uncertain.

The strong, narrow absorption components of the 21-cm line are well suited for attempts to measure the wavelength shift associated with the two directions of circular polarization, and thus to determine the mean magnetic field in the absorbing clouds. For a hydrogen atom in the ground state the orbital angular momentum vanishes, and the Zeeman splitting corresponds to a normal Zeeman triplet, with the Landé g factor set equal to 2. Hence Δv, the difference in frequency between the two components of opposite circular polarization, becomes

$$\Delta v = \frac{eB_\parallel}{2\pi mc} = 2.79 \times 10^6 B_\parallel \text{ Hz} \qquad (2\text{-}60)$$

where $B_\parallel$ denotes the component of the field along the line of sight in gauss. Attempts to measure this very small Δv have yielded [41] an upper limit of about 5×10^{-6} G in various sources.

Visual Lines

Of the 28 identified interstellar lines observed [42] between 3100 and 8000 Å all except for one CN line are absorbed by atoms or molecules in their ground state. Hence the atoms producing interstellar lines in this region of the spectrum are those which have excited states within 4 eV of the ground state. The more abundant atoms which satisfy this restriction are Na I, Ca II, Ca I K I, Ti II, and Fe I; the corresponding interstellar absorption lines have all been observed. However, only the first two are strong enough to permit extensive measurements in a number of stars. Such measurements are normally made in O and B type stars, which show no Na I and not much Ca II in their atmospheric spectra, and which can be observed out to distances of 1 to 2 kpc.

The interstellar Ca II lines at 3968.5 Å (H) and at 3933.7 Å (K)

have been examined with high resolution in the spectra of several hundred early-type stars [43, 44]. Similar but less extensive data have been obtained for the two Na I lines at 5895.9 Å (D_1) and 5890.0 Å (D_2) [45, 44]. As in the 21-cm emission and absorption lines, the observed profiles are complex, indicating a number of components at velocities ranging up to about 100 km/sec, and show clearly the effect of differential galactic rotation. When corrections are made for overlapping of components [46], a dispersion, σ_e, about 8 km/sec in the line of sight seems consistent with the data [39], although an exponential distribution provides a better fit to the observations than does a gaussian [44]. As already noted, the dispersion of velocities for the 21-cm emission components is of this same order, and we shall take 8 km/sec as the dispersion of radial velocities of the interstellar gas clouds within 1 or 2 kpc of the Sun.

The components of high velocity are very much more numerous than would be expected from a Maxwellian distribution with a velocity dispersion of 8 km/sec or from the corresponding exponential distribution. About 18% of the components in Ca II have an absolute value of v, the velocity in the local standard of rest, greater than 15 km/sec, with about 5% greater than 23 km/sec [39]. Corrections for overlapping of the low-velocity components might decrease the percentages by about one-half. These high-velocity components show a basic asymmetry, with negative velocities outnumbering positive ones by about two to one. This evidence suggests that the high-velocity gas is circumstellar, and has been accelerated away from the early-type stars whose spectra show these lines.

Since the stellar distances are known, at least approximately, the average number of Ca II components per kiloparsec can be determined directly from the data. However, measurements [47] on a few stars with much higher resolution show additional components. A theoretical correction for this overlapping for the bulks of the data gives [46] between 8 and 10 components in the Ca II lines per kpc for a line of sight in the galactic plane. A comparable number for neutral hydrogen might, perhaps, be

2. OBSERVATIONS OF THE GAS

consistent with the much more limited data available from the 21-cm absorption components [40]. The gas absorbing a particular velocity component of an interstellar line is frequently referred to as a separate "cloud."

While the normal Ca II clouds seem concentrated to the galactic plane, the clouds producing the high velocity components ($|v| \geq$ 24 km/sec) extend [48] to large z; the mean number of such clouds along a line of sight perpendicular to the galactic plane beyond 750 pc is between 0.25 and 0.5. This number is apparently greater than would be expected from the observed velocity distribution in the plane, if this distribution is assumed isotropic. As is observed for clouds in the galactic plane, negative high velocities are more frequent for clouds far from the galactic plane than are positive ones. Of the five components observed with a velocity exceeding 30 km/sec four have negative velocities; 21-cm emission components of corresponding negative velocities have been observed in all four cases.

The widths of individual absorption components have not been extensively measured. Within several high-velocity components values of about 3 km/sec have been measured [39] for σ_i, the internal r.m.s. dispersion of radial velocities, somewhat greater than the 2 km/sec indicated as the peak of the distribution for the 21-cm emission data (see Fig. 2.2). Interferometric data [47] indicate that some Ca^+ components may have widths corresponding to a value of about 1 km/sec for σ_i.

Equivalent widths of the stronger interstellar lines have been measured and the number of absorbing atoms determined [42]. The ratio of equivalent widths of the two Ca II lines and the corresponding ratio for the two Na I lines, D_2 to D_1, are called "doublet ratios." For each such pair the values of A_{kj} are identical, and the oscillator strengths are in the ratio of the statistical weights of the upper states [see equations (2-22), (2-24), and (2-25)] which equals 2 : 1 for the ^2P term involved. Hence the optical depth in the stronger line is twice that in the weaker. Since $\phi(\Delta v)$ and β must be identical for lines within the same multiplet, the doublet ratio equals $F(2C)/F(C)$, in accordance with equation (2-38).

From Table 2.1 the value of $F(C)$ and of C, the optical depth at the center of the weaker line, can be determined from the measured doublet ratio. The number N_j of absorbing atoms per cm^2 in the line of sight is then determined from equation (2-41). Even if $\phi(\Delta v)$ is given by several partially overlapping Maxwellian profiles, this method will normally give approximately correct results [49] for N_j, though the computed value of β will then exceed that for any one component.

If the doublet ratio is nearly 2, the lines are said to be unsaturated, C is small, and equation (2-36) is applicable, giving a linear relation between the equivalent width, W, and N_j. For stars with unsaturated lines the mean observed density of absorbing atoms is given by [50]

$$n(\text{Ca II}) = n(\text{Na I}) = 2 \times 10^{-9} \text{ cm}^{-3} \qquad (2\text{-}61)$$

As already indicated, the particle density of singly ionized calcium atoms is denoted by $n(\text{Ca II})$ and similarly for other atoms and ions. The density distribution of these atoms appears to be irregular, with $n(\text{Na I})$ and $n(\text{Ca II})$ varying over at least 2 orders of magnitude. For example, the equivalent width of 0.0035 Å for the K line measured in the spectrum of α Vir gives a value of 1.3×10^{-10} cm^{-3} for the mean $n(\text{Ca II})$ over the 90 pc to this star. Equation (2-61) gives a value of $n(\text{Ca II})$ some 15 times larger; since stars with saturated lines were excluded this average represents a mean density somewhat less than the true space average. In the spectra of χ^2Ori, on the other hand, the mean $n(\text{Na I})$ over some 1200 pc is about 3×10^{-8} cm^{-3}, 15 times the mean value; for the high value of $N(\text{Na I})$ in this star, equal to 10^{14} cm^{-2}, the D lines are strongly saturated and $N(\text{Na I})$ was found from the ultraviolet Na I doublet at 3302.9 and 3302.3 Å, with a very low oscillator strength.

If an average value of $N(\text{Ca II})$ is obtained for stars at high latitudes, and divided by the mean $n(\text{Ca II})$ in the galactic plane, the equivalent thickness, 2H, of the Ca$^+$ layer is found [51] to be 240 pc, about the same as for neutral H.

When allowance is made for saturation effects, it appears [42]

2. OBSERVATIONS OF THE GAS

that N(Ca II) and N(Na I) show a definite correlation with the color excess E (see Section 3.2). If an average is taken over all stars, reddened or unreddened [50], the mean density of Na I becomes

$$n(\text{Na I}) = 7 \times 10^{-9} \text{ cm}^{-3} \qquad (2\text{-}62)$$

while for unreddened stars equation (2-61) is applicable. Thus the excess density of neutral sodium associated with interstellar reddening is about 5×10^{-9} cm^{-3}. This analysis requires use of the doublet-ratio method to correct for saturation, and has not been used so extensively for the Ca II lines, since H is generally blended with stellar Hε. Evidently one may conclude that the Na atoms and possibly the Ca II atoms also tend to be concentrated in the clouds responsible for the selective extinction. If F_c is the fraction of space occupied by such dust clouds, then we find for the actual density of neutral Na within these clouds, obtained by adding to the mean density in equation (2-61) the excess density associated with clouds of dust and gas,

$$n(\text{Na I}) = 2 \times 10^{-9} + 5 \times 10^{-9} F_c^{-1} \text{ cm}^{-3} \qquad (2\text{-}63)$$

The relative abundance of neutral sodium and ionized calcium within a particular cloud is observed to depend significantly on the velocity of the cloud relative to the local standard of rest [45]. Table 2.2 gives values of N(Na I)/N(Ca II) averaged over all

TABLE 2.2

Dependence of Sodium-Calcium Ratio on Cloud Velocity

Range of velocity (km/sec)	0–9	10–19	20–39	40 and more
Mean N(Na I)/N(Ca II)	3.7	1.5	0.28	0.45
No. of stars	18	8	7	7

components with radial velocities relative to the local standard of rest in a given range. Evidently Ca II preponderates over Na I in the more rapidly moving clouds, a result indicated also by the greater effective value of β found [52] for saturated Ca II lines than for Na I lines.

Other interstellar lines have not been extensively studied because of their relative weakness. The molecular lines of CH and CH$^+$ observed in stars of spectral types O5 to B1 are roughly correlated with color excess, E, and are likely to be truly interstellar. The lines are all weak and unsaturated, giving mean densities [53]

$$n(\text{CH I}) = 2n(\text{CH II}) = 2 \times 10^{-8} \text{ cm}^{-3} \qquad (2\text{-}64)$$

For stars of spectral type B6 to B9 a few unreddened stars show a strong CH$^+$ line at 4232.6 Å, probably produced by circumstellar molecules. This view is supported by the radial velocity measurements, which show that in all six cases the cloud of molecules observed in the line of sight to a late-type B star is approaching the star, suggesting some physical interrelation.

Of particular interest is the CN line at 3874.0 Å, since this is absorbed by the neutral CN molecule in its first excited rotational level, whose exitation energy is 4.7×10^{-4} eV. The equivalent width of this line in the spectrum of ζ Oph is 0.0034 Å, as compared with 0.0092 Å for the corresponding line from the ground level [54]. The computed ratio of populations of the two levels corresponds to an excitation temperature of 3°K, computed from equations (2-9) and (2-36); the theoretical significance of this result is discussed in Section 4.2.

The unidentified absorption lines [42] are all much broader than the identified ones, with the strongest and widest one, centered at about 4428 Å, showing a width at half maximum depth of about 20 Å [55]. While the equivalent widths of some of these lines are comparable with those of the stronger atomic lines, their origin is still obscure.

2.4 Nonthermal Emission

Measurement of the synchrotron radiation from vaious regions of the Galaxy gives information concerning the particle density of relativistic electrons and the magnetic field in which they gyrate. These electrons are in some ways a relatively minor population in interstellar space, since the primary cosmic ray nucleons outnumber them by about 2 orders of magnitude. However, the

2. OBSERVATIONS OF THE GAS

information on the magnetic field provided by measurements of nonthermal radiation is of major importance. The present section discusses first the observed intensity of this radiation, next the observed flux of energetic electrons and cosmic ray nucleons reaching the Earth, and finally the characteristics deduced for the interstellar magnetic field.

Observed Brightness Temperature

The observed intensity of radio waves from the Galaxy [56, 57] is the result of thermal absorption and emission as well as of nonthermal emission; the former must be taken into account in determining the nonthermal component. According to equation (2-46) the emissivity for free–free transitions, producing the thermal continuum, is independent of frequency, v, when hv/kT is small. Hence I_v is also independent of v if the optical depth is small, and according to equation (2-7) T_b will vary as v^{-2}. For synchrotron radiation, on the other hand, equation (2-52) indicates that T_b varies as $v^{-2-(\gamma-1)/2}$ provided that absorption is negligible; for γ equal to 2.2 the ratio of nonthermal to thermal brightness temperatures varies about as $v^{-0.6}$ Thus the ratio of nonthermal to thermal emission will increase with decreasing frequency. At sufficiently low frequency, however, the continuous absorption by ionized hydrogen becomes important, and T_b will be reduced accordingly if the path length through the gas is sufficiently great.

Typical data for T_b at different values of the galactic latitude, b, are shown in Figure 2.3 [56]. To combine the data for three different wavelengths, values of $T_b/\lambda_m^{2.6}$ are plotted (with λ_m in meters). The coincidence of the three curves at b greater than $4°$ is consistent with the belief that the radiation from intermediate and high latitudes is mostly nonthermal, and indicates that I_v varies as $v^{-0.6}$ for this nonthermal radiation. Most spectral surveys of this radiation show [57] that the exponent of v is between -0.5 and -0.7, corresponding to a value of γ in equation (2-51) between 2.0 and 2.4. This result is generally consistent with

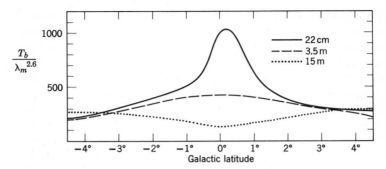

Fig. 2.3. Continuous radio emission at low galactic latitudes. The curves, taken from ref. [56], show the variation of brightness temperature, in °K, with galactic latitude, b, at a galactic longitude of 27°. In accordance with equation (2-52), the values of T_b at each wavelength, λ, are divided by $\lambda_m^{2.6}$., where λ_m is the wavelength in meters.

the observations of cosmic ray electrons reaching the Earth, as summarized below. The peak at $b = 0°$ for 22 cm in Figure 2.3 may be attributed to thermal emission. The assumption that the emission from higher latitudes is mostly nonthermal makes it possible to separate out the thermal component of T_b in the galactic plane; this method was used in determining the r.m.s. proton density in the galactic plane, discussed in Section 2.2. At 3.5 m the brightness temperature of 11,000° observed for radiation in the galactic disc is probably mostly nonthermal, with relatively low thermal obsorption. The dip at 0° for 15 m in Figure 2.3 may be attributed to thermal absorption by the ionized gas.

The absorption of nonthermal radiation at these long wavelengths has been used [58] to measure the amount of ionized hydrogen. At wavelengths between 17 and 140 m the absorption at galactic latitudes greater than 30° is observed to vary as $1/v^2$, in agreement with equation (2-48) for κ_v. If absorption by a uniform layer of ionized gas at 10,000°K is assumed, with an effective half-thickness, H, of 100 pc, the r.m.s. value of n_e is found to be about 0.2 cm^{-3} from the high-latitude data, increasing to about 0.4 cm^{-3} as b decreases to 5°. These results seem entirely consistent

2. OBSERVATIONS OF THE GAS

with the r.m.s. value of 0.45 cm^{-3} found from the thermal emission in the galactic plane [17] (see Section 2.2). At the higher galactic latitudes the extended H II regions which surround O and B stars and which produce much of the thermal radiation in the galactic plane would not be expected to appear. Regions in which H is mostly neutral may also contribute to the observed free–free absorption; if energetic particles maintain an r.m.s. electron density as great as 0.01 cm^{-3}, with T not exceeding 70°K (see Table 4.10), absorption by such H I regions can entirely account for the observed absorption of extragalactic low frequency radio waves.

Earlier work [56] had suggested the existence of a radio "halo" around this and other galaxies, with weak nonthermal emission produced in a spheroidal system some 20,000 pc across. More recent evidence [59] at higher resolution shows a very patchy distribution of surface brightness over the sky and the existence of such a halo is now in serious doubt.

Energetic Particles

The high-energy particles in space which constitute the so-called cosmic rays may play an important part in the observed nonthermal emission, as well as in the thermal balance, the dynamical state, and the evolution of the interstellar medium. Observational information [60, 61] on these "suprathermal" particles, based on extensive measurements from the ground and from space vehicles, is summarized here.

The electron component, though only about a percent of the total cosmic radiation, is of particular importance because of the synchrotron emission which it produces (see Section 2.1). Measurements indicate that the power-law spectrum in equation (2-51) is valid for electrons and positrons with energies between 10^9 and 10^{10} eV, with the following values

$$\gamma = 2; \quad K = 2.7 \times 10^{-15} \text{ particle erg cm}^{-3} \qquad (2\text{-}65)$$

The value of K is applicable if E and dE are both given in ergs.

The fraction of positrons is less than about one-fifth of the total number present.

The protons and other heavy nuclei much outnumber the electrons, and provide most of the energy density in the cosmic radiation. At energies above 10^{10} eV per nucleon the positive-ion spectrum is steeper than for electrons, with γ about equal to 2.6. At somewhat lower energies γ falls to 2, and the total energy density, U_R, of the cosmic radiation does not seem to depend too sensitively on the flux of particles at energies below 10^9 eV, where corrections for the interplanetary magnetic field become large and uncertain. With an error probably less than a factor 2 we have

$$U_R = 1.3 \times 10^{-12} \text{ erg cm}^{-3} \qquad (2\text{-}66)$$

The relative numbers of different nuclei in the cosmic radiation are somewhat similar to the composition of matter in stars and in space, discussed in Chapter 4. The number of energetic protons is about an order of magnitude greater than the corresponding number of He^{++} nuclei, although this comparison is complicated by the different ratios of charge to mass; it is not clear whether the particle densities should be compared for the same energy per nucleon or for the same radius of gyration. For other nuclei the ratios of atomic weight to atomic number are all about 2, and the radii of gyration are about the same if the energies per nucleon are identical. The abundance comparison then gives [60]

$$\frac{n(\text{He})}{n(\text{C}) + n(\text{N}) + n(\text{O})} = 18 \qquad (2\text{-}67)$$

about one-fourth as great as the corresponding ratio for normal matter in the Galaxy (see Table 4.7). This ratio seems to be independent of energy over the entire range from 10^9 to 10^{14} eV. The abundances of the lithium–beryllium–boron group are about a third of the carbon–nitrogen–oxygen group, in marked contrast to the scarcity of these former elements in normal matter. It is usually assumed that these elements are produced by the disruption of the heavier nuclei on collisions with normal matter encountered in space; passage through 3 g/cm² would account

2. OBSERVATIONS OF THE GAS

for the abundances of these light nuclei. The corresponding distance of travel in the galactic disc, where a mean density of 2×10^{-24} g/cm^3 may be assumed (see Table 5.1) is 5×10^5 pc.

Cosmic ray trajectories are dominated by magnetic field effects. The radius of gyration of a 10^{10} eV electron or proton in a field of 3×10^{-6} G is only about 1.1×10^{13} cm, or 0.8 A.U. The high isotropy observed for these energetic particles, to within 0.1%, is consistent with confinement of the particles by a galactic magnetic field. Thus the cosmic-ray particles can travel great distances, gyrating around the lines of force, before escaping from the Galaxy.

The number of energetic particles outside the solar system at energies below 10^9 eV is difficult to determine from the observations because of effects produced by the interplanetary magnetic field. The outwards motion of the interplanetary gas, known as the solar wind, pulls the lines of force with it and tends to sweep away the particles of lower energy. Thus the observed flux of 200-MeV He^{++} nuclei is some 30% greater near Mars than near the Earth [62] and the flux of low-energy particles fluctuates with time. Extrapolation of the power spectrum law, equation (2-51), to energies as low as 10^7 to 10^8 eV gives large but uncertain numbers of particles at these energies. In Section 5.4 we shall see that an appreciable flux of protons with energies of 2×10^6 eV may be produced directly by supernovae. The energy density, U_{Rs}, of such particles, given in equation (5-67), is about an order of magnitude less than the value given in equation (2-66), but the corresponding number of such particles is appreciable; if positive ions of such energies are in fact present with an abundance of this order, they can have an important effect on the equilibrium state of the interstellar gas (see Chapter 4).

Interstellar Magnetic Field

While a number of indirect arguments, with some theoretical uncertainties, have given information on the strength, B, of the interstellar magnetic field, three methods have been used to give a rather direct determination of this quantity. These are: (*a*) the Zeeman effect, especially in the 21-cm H line; (*b*) the synchrotron emission, depending on the density of relativistic electrons; (*c*) the Faraday rotation of plane polarized waves, depending on the total electron density. Method (*a*), already discussed in Section 2.3, gives an upper limit of 5×10^{-6} G for the mean field, with no additional assumptions. Methods (*b*) and (*c*), which both involve nonthermal emission, are described here.

The magnetic field strength may be determined from the brightness temperature, T_b, of synchrotron radiation at a frequency v by use of equation (2-52). We shall use here the value of 11,000° for T_b in the galactic plane at 3.5 m, since, as we have seen above, this radiation is probably mostly nonthermal. Values of K and γ for cosmic-ray electrons may be taken from equation (2-65). It should be noted that a value of 2 for γ gives T_b varying as $v^{-2.5}$, in reasonable agreement with the empirical variation as $v^{-2.6}$ noted in Figure 2.3. If we assume that the density of relativistic electrons is uniform throughout the spiral arms, and take L to be 5 kpc [63], on the picture that the radiation from the disc comes from many spiral arms rather than from the entire disc, equation (2-52) then gives

$$B = 2.9 \times 10^{-5} \text{ G} \qquad (2\text{-}68)$$

At this field the observed radiation at 85 MHz will be produced by electrons whose energy is about 0.8×10^9 eV, if we take $0.3\ v_c$ equal to 8.5×10^7 in equation (2-49).

This field is significantly greater than the upper limit of 5×10^{-6} G obtained from the Zeeman effect. While the synchrotron radiation is produced in a much more extended region than the clouds producing 21-cm absorption, such a large difference in mean field between the two regions is not easily explained. Possibly the synchrotron emission must be explained by emission from expanding supernova remnants, where both the magnetic field and the density of relativistic particles exceed their average values.

Most extragalactic nonthermal radio sources show some plane polarization, with the position angle of the electric vector rotating

2. OBSERVATIONS OF THE GAS

as the frequency is changed. This effect is attributed to Faraday rotation, which results from a slight difference in velocity—about 10^{-7} cm/sec under interstellar conditions—between the two circularly polarized components of a radio wave. For propagation at a frequency v along a mangetic field of intensity B the difference ΔV of phase velocity between waves of opposite circular polarization is given by [7]

$$\frac{\Delta V}{c} = \frac{4\pi n_e e^2}{m} \frac{eB}{mc} \frac{1}{\omega^3} \qquad (2\text{-}69)$$

where ω is the angular frequency and m is again the rest mass of the electron. If plane polarized light is expressed as the sum of two oppositely polarized circular waves, the two circularly polarized components change their relative phase by an angle $2\pi \, \Delta V/c$ in traveling one wavelength, λ. The plane of vibration (defined by the electric vector and the direction of propagation) rotates by half this amount, and the total rotation angle, ψ, over the full path length is given by

$$\psi = \frac{\pi}{\lambda} \int \frac{\Delta V}{c} \, ds = R_m \lambda_m^2 \qquad (2\text{-}70)$$

where the rotation measure, R_m, is defined by

$$R_m = \frac{e^3}{2\pi m^2 c^4} \int n_e B_\parallel \, ds = 8.12 \times 10^5 \int n_e B_\parallel \, ds \text{ pc m}^{-2} \qquad (2\text{-}71)$$

and where λ_m in equation (2-70) is the wavelength measured in meters.

Measures of Faraday rotation in the spectra of some 80 extragalactic radio sources show that R_m increases with decreasing b, and hence a galactic magnetic field of some sort must be responsible. Within about 10° of the galactic plane the observations [64] shown in Figure 2.4 can be fitted roughly by the relationship

$$R_m = 17 \, |\cot b| \, \cos(l - 100°) \qquad (2\text{-}72)$$

Equation (2-72) results if a uniform **B** is assumed in the direction

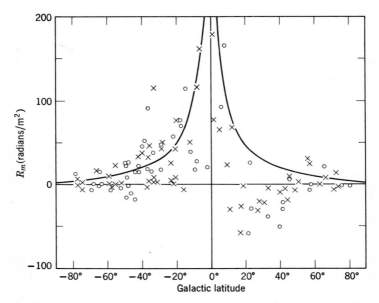

Fig. 2.4. Faraday rotation in extragalactic radio sources. The crosses show values of the rotation measure, R_m, plotted against galactic latitude, b, for stars with galactic longitude, l, between 205° and 355°. The circles show corresponding data for stars with l between 25° and 175°, but plotted with R_m changed in sign. The solid curve represents equation (2-72) with $\cos(l - 100°)$ set equal to unity. The data have been taken from ref. [64].

of l equal to 100°, differing by approximately 30° from the direction of the Orion arm, permeating a uniform layer of gas with a constant n_e. The layer thickness, $2H$, may be taken as 200 pc, about the value found both from the 21-cm data and from the visual absorption lines. We shall here assume that the mean electron density equals 0.06 cm^{-3}, consistent with an r.m.s. electron density of 0.45 cm^{-3}, as found in Section 2.2 from the galactic distribution of ionized hydrogen, and a value of 0.02 for F_i, the fraction of space occupied by H II regions [see equation (2-59)]. On this basis we find

$$B = 3 \times 10^{-6} \text{ G} \qquad (2\text{-}73)$$

2. OBSERVATIONS OF THE GAS

It should be emphasized that the mean n_e is most uncertain. The emission and absorption by H II regions gives information on the mean n_e^2, which is weighted primarily in the localized regions of higher density. A uniform H II layer of density 0.06 cm^{-3}, extending between the clouds of higher density, would produce no measurable Balmer emission and would not much influence the absorption or emission of radio waves, if the gas temperature were about the same as in denser H II regions; the presence of such a layer would double the effective value of n_e, reducing B in equation (2-73) to 1.5×10^{-6} G. Ionization of hydrogen by energetic particles in H I clouds would increase the effective n_e only slightly above the value assumed here (see Table 4.10). On the other hand, if the H II emission and absorption are attributed to smaller, denser clouds, with a resultant decrease in F_i, and if n_e between the clouds is negligible, the mean n_e is appreciably reduced and the value of B would be increased. Also, it will be noted from Figure 2.4 that at higher b the observed values of R_m do not satisfy the simple relation (2-72), and even have the opposite sign, suggesting a reversal in the direction of **B**. These results at higher b may be more influenced by local but large-scale irregularities in the field. Evidently the detailed structure of the magnetic field may be rather complex.

Measures of the polarization of the galactic radio emission [65], presumably synchrotron radiation, are consistent with a magnetic field in the direction of the Orion arm. The magnitude of the polarization is greatest within a 50° band about a great circle perpendicular to the axis of this arm, at l about equal to 70°; the observed plane of vibration is also perpendicular to this axis on the average. As discussed in Section 3.3, the field direction obtained from these results is somewhat discordant with the data both from Faraday rotation [see equation (2-72)] and from optical polarization (see Figure 3.5). The rotation measure R_m observed for this polarized radiation is generally about equal to unity, an order of magnitude less than the values observed for extragalactic sources [see equation (2-72)]. At least some of this reduction can be attributed to the nearly right angle between **B** and the line of sight, which reduces $B_\parallel$ in equation (2-71). Fluctuations in the direction of the plane of vibration [66] over directions of only a few degrees suggest that the direction of the magnetic field changes appreciably over distances smaller than the dimensions of a spiral arm. As shown in Section 3.3 below, a nonuniform component of **B** is indicated also by measures of optical polarization, in qualitative agreement with indications from the radio data.

References

1. E. A. Milne, *Handbuch der Astrophysik*, Springer, Berlin, Chap. 2. Vol, III/1, 1930.
2. S. Chandrasekhar, *Radiative Transfer*, Oxford University Press, London, 1930, Chap. 1.
3. D. H. Menzel, *Selected Papers on Physical Processes In Ionized Plasmas* Dover, New York, 1962; D. H. Menzel and J. G. Baker, *Astrophys. J.*, **88**, 52 (1938).
4. O. Struve and C. T. Elvey, *Astrophys. J.*, **79**, 409 (1934).
5. E. F. M. van der Held, *Z. Physik.*, **70**, 508 (1931); see A. Unsöld, *Physik der Sternatmosphären*, 2nd ed., Springer, Berlin, 1955, Fig. 114, p. 292.
6. C. W. Allen, *Astrophysical Quantities*, Athlone Press, London, 1963.
7. L. Spitzer, *Physics of Fully Ionized Gases, Interscience Tracts on Physics and Astronomy*, No. 3, R. E. Marshak, Ed. 2nd rev. ed., Interscience, New York, 1962.
8. V. L. Ginzburg and S. I. Syrovatskii, *Ann. Rev. Astron. Astrophys.*, **3**, 297 (1965).
9. I. S. Shklovsky, *Cosmic Radio Waves*, translated by R. B. Rodman and C. M. Varsavsky, Harvard University Press, Cambridge, Mass., 1960.
10. H. C. van de Hulst, C. A. Muller and J. H. Oort, *B.A.N.*, **12**, 117, 1954, No. 452.
11. F. J. Kerr and G. Westerhout, in *Stars and Stellar Systems*, Vol. 5, Univ. of Chicago Press, Chicago, 1965, p. 167.
12. F. J. Kerr, in *Stars and Stellar Systems*, Vol. 7, Univ. of Chicago Press, Chicago, 1968, p. 575.
13. C. Heiles, *Astrophys. J., Supp.*, **15**, 97 (1967); No. 136.
14. F. D. Kahn, *Gas Dynamics of Cosmic Clouds (Intern. Astron. Union Symp.* No. 2), North-Holland Publ., Amsterdam, 1958, p. 60.
15. A. E. Lilley, *Astrophys. J.*, **121**, 559 (1955).
16. R. D. Davies, *Monthly Notices, Roy. Astron. Soc.* **116**, 443 (1956).
17. G. Westerhout, *B.A.N.*, **13**, 201 (1957), No. 475.
18. K. Takakubo and H. van Woerden, *B.A.N.*, **18**, 488 (1967).
19. J. H. Oort, *B.A.N.*, **18**, 421 (1966).

2. OBSERVATIONS OF THE GAS

20. T. K. Menon, *Astrophys. J.*, **127**, 28 (1958).
21. C. M. Wade, *Third Symp. on Cosmical Gas Dynamics* (*Intern. Astron. Union Symp.* No. 8), *Rev. Mod. Phys.*, **30**, 946 (1958).
22. K. W. Riegel, *Astrophys. J.*, **148**, 87 (1967).
23. S. R. Pottasch, *Vistas Astron.*, **6**, 149 (1965).
24. H. M. Johnson, in *Stars and Stellar Systems*, Vol. 7, Univ. of Chicago Press, Chicago, 1968, p. 65.
25. H. M. Johnson, *Astrophys. J.*, **118**, 370 (1953).
26. L. H. Aller and W. Liller, in *Stars and Stellar Systems*, Vol. 7, Univ. of Chicago Press, Chicago, 1967, p. 483.
27. D. Osterbrock and E. Flather, *Astrophys. J.*, **129**, 26 (1959).
28. C. R. O'Dell, *Astrophys. J.*, **143**, 168 (1965).
29. G. R. Burbidge, R. J. Gould and S. R. Pottasch, *Astrophys. J.*, **138**, 945, 1963.
30. G. Courtes, P. Cruvellier and S. R. Pottasch, *Ann. d'Astrophys.*, **25**, 214 (1962).
31. O. C. Wilson, G. Münch, E. M. Flather and M. F. Coffeen, *Astrophys. J., Suppl.*, **4**, 199 (1959), No. 40.
32. R. Minkowski, in *Stars and Stellar Systems*, Vol. 7, Univ. of Chicago Press, Chicago, 1968, p. 623.
33. G. Westerhout, *B.A.N.*, **14**, 215 (1958), No. 488.
34. F. T. Haddock, *Radio Astronomy* (*Intern. Astron. Union Symp.* No. 4), Cambridge University Press, 1957, p. 192.
35. N. S. Kardashev, *Astron. Zh.*, **36**, 838, 1959; *Soviet Astron. A.J.*, **3**, 813 (1959).
36. P. G. Mezger and B. Höglund, *Astrophys. J.*, **147**, 490 (1967).
37. L. Goldberg, *Astrophys. J.*, **144**, 1225 (1966).
38. B. J. Robinson and R. McGee, *Ann. Rev. Astron. Astrophys.*, **5**, 183 (1967).
39. L. Spitzer, in *Stars and Stellar Systems*, Vol. 7, Univ. of Chicago Press, Chicago, 1968, p. 1.
40. B. G. Clark, *Astrophys. J.*, **142**, 1398 (1965).
41. G. L. Verschuur, *Radio Astronomy and the Galactic System* (*Intern. Astron. Union Symp.* No. 31), Academic Press, London, 1967, p. 385.
42. G. Münch, in *Stars and Stellar Systems*, Vol. 7, Univ. of Chicago Press, Chicago, 1968, p. 365.
43. W. S. Adams, *Astrophys. J.*, **109**, 354 (1949).
44. G. Münch, *Astrophys. J.*, **125**, 42 (1957).
45. P. McR. Routly and L. Spitzer, *Astrophys. J.*, **115**, 227 (1952).
46. A. Blaauw, *B.A.N.*, **11**, 405 (1952), No. 433.
47. L. M. Hobbs, *Astrophys. J.*, **142**, 160 (1965).
48. G. Münch and H. Zirin, *Astrophys. J.*, **133**, 11 (1961).
49. B. Strömgren, *Astrophys. J.*, **108**, 242 (1948).
50. L. Spitzer, *Astrophys. J.*, **108**, 276 (1948).

51. P. J. Van Rhijn, *Publ. Kapteyn Astron. Lab.*, No. 50, Groningen, 1946.
52. P. W. Merrill and O. C. Wilson, *Astrophys. J.*, **86**, 44 (1937).
53. D. R. Bates and L. Spitzer, *Astrophys. J.*, **113**, 441 (1951).
54. G. B. Field and J. L. Hitchcock, *Astrophys. J.*, **146**, 1 (1966).
55. G. H. Herbig, *Z. Astrophys.*, **64**, 512 (1967).
56. J. L. Pawsey, in *Stars and Stellar Systems*, Vol. 5, Univ. of Chicago Press, Chicago, 1965, p. 219.
57. B. Y. Mills, *Ann. Rev. Astron. Astrophys.*, **2**, 185 (1964).
58. G. R. A. Ellis and P. A. Hamilton, *Astrophys. J.*, **146**, 78 (1966).
59. J. E. Baldwin, *Radio Astronomy and the Galactic System* (*Intern. Astron. Union Symp.* No. 31), Academic Press, London, 1967, p. 337.
60. W. R. Webber, *Handbuch der Physik*, **46/2**, 181, 1967.
61. V. L. Ginzburg and S. I. Syrovatskij, *Radio Astronomy and the Galactic System* (*Intern. Astron. Union Symp.* No. 31), Academic Press, London, 1967, p. 411.
62. J. J. O'Gallagher and J. A. Simpson, *Astrophys. J.*, **147**, 819 (1967).
63. B. Y. Mills, *Paris Symp. Radio Astron.* (*Intern. Astron. Union Symp.* No. 9), Stanford Univ. Press, 1959, p. 431.
64. R. D. Davies, *Nature*, **218**, 435 (1968).
65. F. F. Gardner and J. B. Whiteoak, *Ann. Rev. Astron. Astrophys.*, **4**, 245 (1966).
66. H. C. van de Hulst, *Ann. Rev. Astron. Astrophys.*, **5**, 167 (1967).

Chapter 3

Observations of the Grains

3.1 Theory of Absorption and Scattering by Grains

As radiation passes through a region of space containing solid particles, or grains, the photons can be either absorbed, with conversion of radiant energy into heat, or scattered, with a change in the photon's direction of travel. The sum of these two processes is called "extinction," since a collimated beam of light can be extinguished by either process. This terminology differs somewhat from that in the previous chapter; in the discussion of atomic and molecular processes it is customary to describe as absorption any photon-induced excitation, even though a prompt deexcitation, with photon reemission, leads essentially to scattering. The basic equations are the same, and we shall use equation (2-3) to determine the intensity of radiation reaching the Earth either from a star beyond a dust cloud or from the dust cloud itself. However, the coefficient κ_ν entering into the optical depth [equation (2-2)] will here be called an "extinction coefficient."

First we consider the extinction which grains produce in the light from a distant star. For observations of starlight $I_\nu(0)$ in equation (2-3) is the true intensity of light from the star; the contribution to I_ν within a stellar image resulting from scattering or emission by interstellar grains is usually negligible, and the second term on the right-hand side of equation (2-3) may be ignored.

Moreover, it is generally the flux, $\mathscr{F}_v$, obtained by integrating I_v over the solid angle of the star, that is actually observed. If we express the optical thickness, τ_v, in terms of the coefficient, s_v, per particle, in accordance with equation (2-20), equation (2-3) yields for A_v, the interstellar extinction in magnitudes

$$A_v = -2.5 \log \frac{\mathscr{F}_v}{\mathscr{F}_v(0)} = 1.086 \, N_g Q_e \sigma_g \qquad (3\text{-}1)$$

where $\mathscr{F}_v(0)$ is the stellar flux at the Earth in the absence of extinction, N_g is the number of grains per cm^2 along the line of sight from the Earth to the star, σ_g is the geometrical cross section of a single grain, and Q_e is the "extinction efficiency factor," defined in terms of the optical cross section, s_v, by the relationship

$$Q_e = \frac{s_v}{\sigma_g} \qquad (3\text{-}2)$$

The extinction efficiency of the grains and the resulting extinction both depend on v. These equations are based on the assumption that all the grains are identical. In the actual case equation (3-1) must be integrated over all the parameters characterizing the grains, including chemical composition, size, shape, and orientation.

Next we consider scattering of light by grains over an appreciable area of the sky, producing a diffuse source of illumination. In this case it is the intensity, I_v, rather than the flux, $\mathscr{F}_v$, that is observed, and the emissivity, j_v, makes the interesting contribution in equation (2-3). We define a "scattering efficiency factor" Q_s, so that a fraction Q_s/Q_e of the light extinguished is scattered rather than absorbed; this fraction is called the "albedo." We may write

$$j_v(\boldsymbol{\kappa}) = n_g Q_s \sigma_g \int I_v(\boldsymbol{\kappa}')F(\boldsymbol{\kappa} - \boldsymbol{\kappa}') \, d\omega' \qquad (3\text{-}3)$$

where $\boldsymbol{\kappa}'$ and $\boldsymbol{\kappa}$ are vectors indicating the directions of the incident and scattered photons, $d\omega'$ is an interval of solid angle about $\boldsymbol{\kappa}'$, and $F(\boldsymbol{\kappa} - \boldsymbol{\kappa}')$ is a phase function, which we may assume depends

3. OBSERVATIONS OF THE GRAINS

on the angle, ϕ, between κ and κ'; the integral of F over all $d\omega'$ is unity.

The difference $Q_e - Q_s$ is the efficiency factor for true absorption, which we denote by Q_a. The total energy absorbed by the grains per sec per cm^3 and reemitted as thermal radiation is given by integrating $n_g Q_a \sigma_g I_\nu(\kappa') \, d\nu \, d\omega$ over all $d\nu$ and all directions κ'.

To interpret observational data on interstellar extinction and scattering, the values of Q_e and Q_s for different types of grains must be known. Extensive theoretical calculations of these efficiencies have been carried through for spherical particles, and additional information is available for cylinders and for spheroids [1, 2]. This knowledge is summarized below, together with what little is known about other more complex types of grains.

Spheres

The Mie theory of scattering and absorption by spheres with a complex index of refraction, m, has been applied numerically to a wide variety of cases [1]. Here we give a few typical results and discuss in particular the asymptotic behavior for spheres whose radius, a, is much larger or much smaller than the wavelength. It is customary to express the size of the sphere in terms of the dimensionless parameter, x, defined by

$$x = \frac{2\pi a}{\lambda} \tag{3-4}$$

In Figure 3.1 are shown values of Q_e for spheres with four different values for m, the index of refraction: (a) $m = \infty$, corresponding to an infinite dielectric constant; (b) $m = 1.33$, corresponding to ice particles; (c) $m = 1.33 - 0.09i$, corresponding to ice with absorbing impurities, or "dirty ice"; (d) $m = 1.27 - 1.37i$, corresponding to spheres of iron. The horizontal scale is x; as m approaches unity, the curves for Q_e shift to the right by an amount proportional to $1/(m - 1)$. For small x, the scattering efficiency factor, Q_s, becomes very small, and, if mx is also small, the theory of Rayleigh scattering gives

$$Q_s = \frac{8}{3} x^4 \left| \frac{m^2 - 1}{m^2 + 2} \right|^2 \tag{3-5}$$

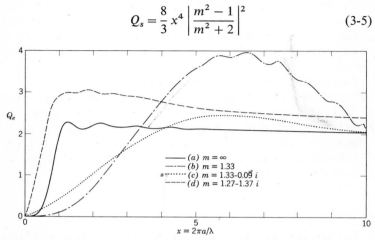

Fig. 3.1. Extinction cross section for spheres, The curves show Q_e, the ratio of the extinction cross section to the geometrical cross section, for spheres with different values of m, the index of refraction; x is the ratio of the sphere's circumference to the wavelength, λ, of the incident light. All values have been obtained from ref. [1]; those for cases (a), (b), and (d) are based on exact computations, but in case (c) the curve is found from an approximate theory, valid for small $m - 1$.

When m is purely real (spheres a and b in Figure 3.1), Q_a vanishes, but for the iron sphere much of the extinction is due to absorption. Introduction of a small imaginary part in the refractive index of the ice sphere does not change Q_s appreciably, but for small x yields a Q_a proportional to the particle radius (i.e., proportional to x with a fixed λ), exactly as with the iron particle; this accounts for the increased Q_e for small x for the dirty ice sphere. The large-scale oscillation of Q_s with changing x for the sphere of ice, which is due to interference between the transmitted and diffracted radiation, is evidently reduced when absorption within the sphere is assumed.

3. OBSERVATIONS OF THE GRAINS

Figure 3.1 shows that as x increases, Q_e approaches an asymptotic value of about 2, both for absorbing and for dielectric spheres. Thus twice as much energy is removed from the beam as is actually striking a large sphere. The reason for this behavior is that, in accordance with Babinet's principle, the diffraction pattern due to an obstacle is identical with that of an aperture of the same cross section, and thus contains the same energy as is striking the obstacle. The diffracted energy can be regarded as removed from the incident beam if two requirements are met. Firstly, energy measurements must be made at such great distances behind the obstacle that the shadow has been "washed out," and the Fraunhofer pattern is applicable. Secondly, the diffraction angles must exceed both the initial angular spread of the incident beam and the angular resolution of the detection apparatus.

Evidently this diffracted light accounts for at least half of the energy scattered by a large sphere, with the rays passing through the sphere accounting for the other half if the material is non-absorbing. We evaluate the contribution of this diffracted radiation to the phase function F, which occurs in equation (3-3) for the scattered light. If a spherical grain is considered, F is a function $F(\phi)$ of the angle ϕ between the incident light and the direction of the scattered or diffracted light; according to equation (3-3) $F(\phi) \, d\omega$ is the fraction of the scattered light energy which at great distances is traveling within a small cone, of solid angle $d\omega$, at an angle ϕ to the direction of the incident beam of light. The diffracted component of this radiation follows closely the Fraunhofer diffraction pattern for a circular aperture with a radius, a, equal to the radius of the sphere, and we have the familiar result,

$$F(\phi) = \frac{J_1^2(x \sin \phi)}{\pi \sin^2 \phi} \quad (3\text{-}6)$$

The normalization is exact only in the limit as x becomes large. In this equation $J_1(u)$ denotes a Bessel function of order 1, which approaches $u/2$ as u falls below unity. If we denote by ϕ_1 the value of ϕ at which $F(\phi)$ falls to half its central value, we obtain, if ϕ_1 is sufficiently small so that $\sin \phi_1$ may be replaced by ϕ_1,

$$\phi_1 = \frac{1.617}{x} = 0.257 \frac{\lambda}{a} \tag{3-7}$$

The total scattered light includes radiation which has been refracted through the sphere, and which is not included in equation (3-6). However exact computations [1] for a sphere with m equal to 1.33 show that equation (3-7) gives ϕ_1 rather accurately for the total scattered light when x is between 2 and 5.

Cylinders and Spheroids

The values of the efficiency factors for spheroids and cylinders tend to behave in about the same way as for spheres. The interesting feature concerning these more complex particles is that Q_e may differ depending on the direction of polarization; we denote by Q_{eE} and Q_{eH} the values of Q_e when the axis of symmetry of the cylinder or spheroid is parallel to the electric vector and to the magnetic vector, respectively, of the incident radiation.

Exact computations have been carried out for infinite cylinders [1]. Values of Q_e are shown [2] in Figure 3.2 for m equal to 1.33 in the simple case where the wave front is parallel to the cylinder axis. Evidently $Q_{eE} - Q_{eH}$ is relatively constant with increasing x, but falls off for sufficiently large x. For small x, the cylinders behave as small antennas, and Q_{eE}/Q_{eH} approaches infinity as x goes to zero. Addition of some absorption (a small imaginary component of m) changes Q_{eE} and Q_{eH} in much the same way as Q_e is changed for a sphere (compare curves (*b*) and (*c*) in Figure 3.1); however, $Q_{eE} - Q_{eH}$ is not much altered.

For spheroids [2] the exact computations have been carried through only for particles small compared to the wavelength. For such "Rayleigh-Gans" particles both Q_{eE} and Q_{eH} vary as x^4, but their ratio is constant with x, and equals the square of the ratio of static polarizabilities in the two directions. The optical properties of finite spheroids have been measured in the laboratory at

3. OBSERVATIONS OF THE GRAINS

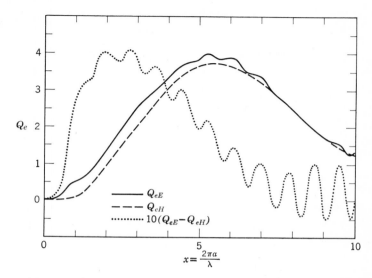

Fig. 3.2. Extinction cross section for cylinders. The curves, taken from ref. [2], show Q_{eE} and Q_{eH}, the extinction efficiency factors for cylinders whose axes are parallel to **E** and **H**, respectively, in the incident radiation. The direction of propagation is perpendicular to the cylinder axis; a is the cylinder radius, and the index of refraction within the cylinder is taken to be 1.33.

microwave frequencies with model spheroids scaled to give the correct $2\pi a/\lambda$, where a is the semiminor axis. For a prolate spheroid with its semimajor axis, b, equal to $2a$, the measured values [2] of Q_{eE} and Q_{eH} are not very different from the theoretical values for an infinite cylinder of the same index of refraction, with a radius equal to a. In particular, the values of $Q_{eE} - Q_{eH}$ for such spheroids are generally at least half or more of the corresponding values for the cylinders.

The polarizing effectiveness of cylindrical particles with values of m, the index of refraction, equal to 1.25 and 1.33 is summarized [2] in Table 3.1. The quantity r_p is the ratio of $Q_{eE} - Q_{eH}$ to $\overline{Q_e}$, where $\overline{Q_e}$ is defined as $(Q_{eE} + Q_{eH})/2$. As may be seen

TABLE 3.1

Extinction Efficiencies for Infinite Cylinders

	$m = 1.25$		$m = 1.33$	
x	$\overline{Q_e}$	r_p	$\overline{Q_e}$	r_p
1	0.164	0.99	0.285	1.05
2	0.645	0.38	1.11	0.37
3	1.37	0.19	2.20	0.17
4	2.19	0.110	3.18	0.085
5	2.94	0.067	3.76	0.049

from equations (3-1) and (3-2), combined with equation (3-11), r_p is the ratio of polarization to extinction in the ideal case where cylindrical grains are all aligned parallel to each other and perpendicular to the line of sight.

Complex Particles

In addition to the simple uniform classical particles discussed above, a number of more complex types of grains have been considered. Graphite flakes [3, 4] are characterized by anisotropic optical properties, since for an electric field parallel to the plane of the flake the material is conducting, while for **E** perpendicular to the plane the material is dielectric. Oriented graphite particles small compared to the wavelength will therefore yield a large difference in Q_e between the two directions of polarization. To investigate the optical properties of such particles computations have been made [4] for spheres with an isotropic refraction index, m, taking for m the measured values with **E** parallel to the conducting planes. Similar computations have also been made [4] for graphite spheres surrounded with mantles of ice, for which the ratio Q_s/Q_e is more nearly unity. The resultant extinction cross sections, both for pure graphite and for the compound spheres with ice mantles, all show a pronounced maximum at about 2200 Å, where the measured value of m is greatest. Compound particles composed of metal cores with dielectric mantles have also been investigated [2].

3. OBSERVATIONS OF THE GRAINS

Particles which scatter radiation by quantum mechanical transitions up to unfilled electronic energy bands have been proposed [5]. In these hypothetical "free radicals," similar to unsaturated organic compounds, the value of Q_e might be as great as unity for $2\pi a/\lambda$ as low as 10^{-2}. Such particles might have a Q_s/Q_e nearly equal to unity; in any case the energy should be reradiated nearly isotropically, since the particle size is so much smaller than the wavelength. On the other hand, the extinction properties of such particles might well be sufficiently anisotropic to produce some polarization, if the particles were aligned. Definite evidence is lacking on the physical possibilities of such small scattering particles.

3.2 Extinction and Scattering by the Grains

Many measurements of light from stars and from the sky are influenced by the interstellar solid particles, or grains. The most complete information about the nature of the grains has been obtained from the increase of extinction with decreasing wavelength, called "selective extinction," or "reddening" of the light from distant stars. Determinations of the total extinction have been less definite. Measurements of the starlight scattered or reflected by grains can in principle supplement the selective extinction data in determining the size and composition of the grains. The present section discusses in turn the results obtained in each of these three areas.

Selective Extinction

The goal of selective extinction studies is to measure how A_v, the extinction in magnitudes, varies with frequency, v, without necessarily determining the magnitude of A_v. To achieve this, two stars whose spectral type and luminosity class are known to be identical from measurements of their spectral lines are compared, and the magnitude difference $\Delta m(\lambda)$ between them is measured at two wavelengths, λ_1 and λ_2. Measurements are usually made on stars of spectral type O and B, partly because they are far away and show substantial amounts of extinction, and partly because the shape of the intrinsic continuous spectrum for the hotter stars does not depend very sensitively on spectral type for wavelengths in the visible. Since the emitted spectra of the two stars are presumably identical, each value of $\Delta m(\lambda)$ is the sum of two terms, one varying as the logarithm of the distance ratio, one depending on the difference in A_v between the two stars. If now we take the difference $\Delta m(\lambda_1) - \Delta m(\lambda_2)$, the terms depending on the distances cancel out, and we have

$$\Delta m(\lambda_1) - \Delta m(\lambda_2) = \Delta(A_{v_1} - A_{v_2}) \qquad (3\text{-}8)$$

where Δ denotes the difference between the two stars. If for one of the two stars A_v is thought to be zero, either because the star is very close, or because of agreement between the measured continuous spectrum and theoretical calculations, then equation (3-8) gives directly $A_{v_1} - A_{v_2}$ for the other star, a quantity which may be denoted by $E(\lambda_1, \lambda_2)$. In practice the shape of the intrinsic or unreddened spectrum is usually determined separately for a group of relatively close stars. Since there may be some uncertainty in this determination, a small zero-point error in $E(\lambda_1, \lambda_2)$ is usually a possibility. If λ_1 and λ_2 are taken to be 4350 and 5550 Å, about equal to the mean wavelengths for the blue (B) and visual (V) photometric bands [6], respectively, then $E(\lambda_1, \lambda_2)$ becomes the standard color excess $E_{B\text{-}V}$.

Measures of $E(\lambda_1, \lambda_2)$ have been used in two ways. With λ_1 and λ_2 fixed, observations of many stars indicate how the color excess, and presumably the number of grains in the line of sight, vary with direction and distance. With λ_2 fixed, measurements have been made with variable λ_1 to determine the variation of selective extinction with wavelength.

Photoelectric measures of some 1300 B stars of known spectral type have given [7] an extensive catalog of color excesses. Approximate distances are known for these stars, since the absolute magnitude can be estimated from the spectra, and the total extinction may be set equal to $3E_{B\text{-}V}$ (see the discussion below). These data indicate that the color excess can have quite different values for

3. OBSERVATIONS OF THE GRAINS

stars at the same distance, and as a result a mean reddening coefficient in magnitudes per kpc cannot be used to predict E_{B-V} at a distance L. However, for computing the mean amount of dust present in the galactic plane, within a distance of 1000 pc from the Sun, the following general average over all directions [8] (see Table 3.3) is useful.

$$\frac{E_{B-V}}{L} = 0.61 \text{ mag/kpc} \qquad (3-9)$$

where color excesses on the C_1 system have been multiplied by 2.06 to convert them [9] to the B-V system. It is possible that this mean value should be increased somewhat because of observational selection, the more highly reddened stars being too faint to observe. The color excess in the direction of the galactic poles, as measured from the reddening of galaxies and globular clusters, averages [10] about 0.055 mag. Division of twice this value by the mean color excess per unit distance in equation (3-9) gives an effective thickness 2H of 180 pc for the absorbing layer, somewhat smaller than the 220 and 240 pc found for H and Ca II atoms, respectively, in Sections 2.2 and 2.3. The irregular distribution of the interstellar material makes all these values about equally uncertain.

The variation of $E(\lambda, \lambda_2)$ with λ has been measured [4, 11] both with wideband photometry and with spectrophotometry. To combine data from different stars, with different numbers of absorbing grains, it is customary [6] to normalize $E(\lambda, \lambda_2)$, dividing by $E(\lambda_1, \lambda_2)$ where λ_1 is some fixed wavelength. Typical results are shown in Figure 3.3, where $E(\lambda, V)/E_{B-V}$ is plotted against $1/\lambda$, with λ varying from 3000 Å to 2 μ. In this normalization λ_1 and λ_2 are again taken to be the wavelengths corresponding to the B and V bands [6]. Observed points [12] are averages for several pairs of stars.

Because of equations (3-1) and (3-8) the curve in Figure 3.3 is directly comparable, apart from scale factors and zero points, with the curves for Q_e shown earlier in this chapter. The solid line in Figure 3.3 is a theoretical curve [13] for ice grains ($m = 1.33$) of two radii, 2.9×10^{-5} cm and twice this radius, with the particle

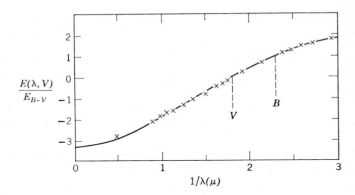

Fig. 3.3. Dependence of selective extinction on wavelength. The ratio of $E(\lambda, \lambda_2)$ to $E(\lambda_1, \lambda_2)$ is plotted against the reciprocal wavelength in microns, where λ_1 and λ_2 are the wavelengths of the B and V bands, respectively, and $E(\lambda_1, \lambda_2)$ is accordingly the color excess E_{B-V}. The points are taken from ref. [12]; the solid line is a theoretical curve [13] for a mixture of ice grains with two radii, 0.29 and 0.58 μ. The effective wavelengths corresponding to the V and B bands are indicated by the dashed lines

density of the larger grains 1/40 of that for the smaller, corresponding to 1/5 the mass density. This curve gives a reasonable fit to the observations. About the same curve has been obtained [4] for graphite particles; for iron particles also [14] a reasonable fit is possible. The radii of the graphite or iron spheres required are about 5×10^{-6} cm.

More complete photometric data [11] on selective extinction between 3000 Å and 3.5 μ with some data at wavelengths 5 and, 10 μ, indicate that in some regions $E(\lambda, V)/E_{B-V}$ decreases abruptly at the longest wavelengths, suggesting that $-E(\infty, V)/E_{B-V}$, may be twice as great as the value 3 indicated in Figure 3.3. Such a wavelength variation of $E(\lambda, V)$ could be produced by particles with radii exceeding 10^{-4} cm. In addition, preliminary observations of $E(\lambda, V)/E_{B-V}$ at wavelengths between 1300 and 2600 Å [15] lie significantly above the theoretical curve in Figure 3.3, indicating an appreciable number of grains with smaller radii.

3. OBSERVATIONS OF THE GRAINS

The mass density of interstellar grains required to account for the mean color excess given in equation (3-9) may be computed for each of the models used to fit the shape of the extinction curve in Figure 3.3. For grains of ice, assumed to have enough impurities to increase their internal density to 1.1 g/cm^3 the space density is about 1.3×10^{-26} g/cm^3. About the same density is obtained if the theoretical size distribution curve referred to in Section 4.6 is assumed. The space densities required for either graphite or iron particles have about these same values. For free radicals, however, the space density would be less by about two orders of magnitude. The total cross-sectional area per unit volume, equal to $n_g \sigma_g$, is much the same on all four models, and equals about 3.2×10^{-22} cm^{-1}. If an appreciable number of larger particles is present, the overall density and cross-sectional area of the grains will be increased correspondingly. Uncertainties in the number of smaller particles are not likely to affect the overall grain density appreciably, since the geometrical cross section per unit mass carries inversely as the grain radius.

General Extinction

The general extinction is measured by A_V, the value of the extinction A_v at the V wavelength, 5550 Å. The quantity R_V is defined as

$$R_V = \frac{A_V}{E_{B-V}} \tag{3-10}$$

In Figure 3.3 R_V equals $-E(\infty, V)/E_{B-V}$, i.e., the normalized selective extinction at zero frequency, at which the extinction is assumed to vanish; a value of about 3 seems indicated by these data. The selective extinction data at infrared wavelengths [11] referred to above indicate that R_V may vary from 3 to 6, depending on the particular region being measured.

If A_V can be measured entirely independently, the resultant value of R_V can be used to set a limit on the density of large particles, producing "grey extinction" in visible light. If the sizes of such particles exceeded λ for the longest wavelength used in the measures of selective absorption, their presence would not affect the shape of the observed selective extinction curve, but would increase R_V above the value found from this curve.

Unfortunately, precise measures of A_V are difficult, since it is not easy to determine what the brightness of an object would be in the absence of absorption. In principle the extinction of light from external galaxies as a function of the galactic latitude, b, can provide such a measurement, since the optical thickness τ along the line of sight out of the Galaxy (about equal to the extinction) varies as $\tau_0 \csc b$, where τ_0 is the optical thickness toward the galactic poles. The observed number of galaxies per unit solid angle brighter than some apparent magnitude, m, denoted by $N(m)$, increases with m. As a result of the extinction this function will be shifted to larger m by an amount proportional to $\csc b$, and measurement of this shift yields, in principle, a determination of τ_0. The visual extinction determined in this way toward the galactic poles (equal to 1.086 τ_0) lies between 0.25 and 0.4 mag [16], the value depending on the corrections made to the photographic photometry used. The color excess of 0.058 mag observed in globular clusters and galaxies [10] in the direction of the galactic poles then gives R_V between about 4 and 7.

Within a galactic cluster the spottiness of extinction changes both A_V and E_{B-V} from star to star. While A_V cannot be measured directly, the deviations of the stars from a single sequence on the Hertzprung-Russell diagram can give the deviations of both A_V and E_{B-V} from the mean. The correlation between these deviations can then give R_V directly. Values of R_V determined in this way seem to show [11] a general agreement with those obtained from the selective extinction curves out to wavelengths of 5 μ or more, ranging from 3 to 6; however, the precision of these determinations is low. Other methods give about the same result. We may infer that the amount of grey extinction in most interstellar clouds is somewhat uncertain, but seems unlikely to correspond to R_V greater than 7. If this upper limit were to hold in the galactic plane generally, instead of the more generally accepted value of about 3, the visual extinction found from equation (3-9) would have the

3. OBSERVATIONS OF THE GRAINS

rather large value of 4.3 mag/kpc; the density of interstellar grains would be increased by a factor between 2 and 4, depending on the diameters of the larger particles assumed.

Direct measures of A_V from star counts are not sufficiently precise to give information on the nature and composition of the grains, but give data on the spatial distribution of dust. Important evidence obtained in this way [17] on the distribution of absorbing clouds is summarized in Section 3.4.

Scattering

Analysis of starlight scattered from dust near the galactic plane, measured at low galactic latitudes, has led to important but tentative conclusions concerning the grains. The emissivity for this scattered light in the direction κ is given in equation (3-3), where $I_\nu(\kappa')$ in the integral is the intensity of starlight in the direction κ'. The integral of $n_g \, Q_e \, \sigma_g \, ds$ along the line of sight is the optical thickness through the Galaxy in the direction κ, which can be estimated from counts of external galaxies. Thus the chief unknowns in equation (3-3) are Q_s/Q_e, which is the albedo of the grains, and the phase function $F(\kappa - \kappa')$. The intensity of starlight, $I_\nu(\kappa')$, has a sharp maximum for κ' nearly parallel to the galactic plane. The measurements of the scattered light intensity, $I_\nu(\kappa)$, are mostly at low galactic latitudes, and for κ nearly parallel to the galactic plane j_ν will be greatest if $F(\kappa - \kappa')$ corresponds to a sharply forward-throwing phase function; i.e., if x in equation (3-6) is large and ϕ_1 consequently small. If either I_ν or F were isotropic the integral in equation (3-3) would equal the mean I_ν, averaged over solid angle, but for a forward-throwing phase function the integral can approach the peak value of $I_\nu(\kappa')$ for κ sufficiently close to the galactic plane.

Measurement of the scattered light is appreciably complicated by corrections for light from many faint stars and by other observational uncertainties. Results indicate [18] that this scattered light within some 10° of the galactic plane is relatively intense, much brighter than would be expected if the grains were iron particles of small x, characterized by a low Q_s/Q_e and by a nearly isotropic phase function for the scattered radiation. Even if Q_s/Q_e is taken to be unity, the observations require that the phase function be moderately forward-throwing, with an effective half angle, ϕ_1, of about 0.4 radian. If this conclusion is accepted, both graphite particles and free radicals may be excluded as the dominant constituents of the medium, since both of these particles, characterized by small x, will scatter nearly isotropically. However, for ice grains with a radius of 2.9×10^{-5} cm, the value found above for the dominant component, x equals 4.2 at 4300 Å, and the value of ϕ_1 found from equation (3-7) is 0.37 radian, about equal to the value suggested by the observations. The tentative conclusion is that the interstellar grains are primarily dielectric. Confirmation of this observational result would be most important.

Measures of light scattered from dust in individual nebulae have shown [19] that with a few significant exceptions the ratio of dust to gas is about the same in H I and H II regions. More detailed studies of reflection nebulae [20] have not yielded much information on the properties of the scattering particles. In particular, geometrical uncertainties on the location of reflection nebulae relative to the star illuminating them have made it difficult to draw conclusions concerning Q_s/Q_e or the phase function.

3.3 Polarization

Since the initial discovery of plane polarization [21, 22] in reddened stars, measurements by many observers have been made on some 4000 stars [23]. Each measurement involves a determination of the flux $\mathscr{F}$ in some wavelength band as a function of the direction of the electric vector in the radiation. If $\mathscr{F}$ varies from a maximum value, $\mathscr{F}_{max}$, to a minimum, $\mathscr{F}_{min}$, as the direction of the electric vector passed by the analyzing photometer is rotated, the polarization in stellar magnitudes is given by

$$p = 2.5 \log \frac{\mathscr{F}_{max}}{\mathscr{F}_{min}} \qquad (3\text{-}11)$$

3. OBSERVATIONS OF THE GRAINS

Observed values of p rarely exceed 0.15 mag. The direction of the electric vector for which $\mathscr{F}$ equals $\mathscr{F}_{max}$ tends to be parallel to the galactic plane; the plane defined by this direction of the electric vector and the direction of propagation is called the "plane of vibration." The subsequent discussion treats the variation of the observed polarization with color excess, with wavelength and with galactic coordinates.

Dependence on Color Excess

In general, unreddened stars show no polarization. However, stars with large color excess may or may not show a large p. The observed values of p/E_{B-V} show [24] a large scatter, with a maximum value given by

$$\left(\frac{p}{E_{B-V}}\right)_{max} = 0.19 \tag{3-12}$$

The mean value of p/E_{B-V} for a group of some 1300 O and B stars is about half this maximum.

The strong correlation observed between reddening and the maximum polarization indicates that the grains are responsible for the polarization. This result requires (*a*) that the grains are anisotropic, with the extinction efficiency factor, Q_e, varying from a maximum value, Q_{max}, to a minimum value, Q_{min}, for different directions of the electric vector, and (*b*) that the grains are somehow aligned so that on the average Q_e tends to have its maximum value for the same orientation of the electric vector. If the grains are identical and perfectly aligned, then the radiation is most heavily absorbed if the electric vector is oriented so that Q_e equals Q_{max}. The radiation will then appear partially plane polarized with the plane of vibration in the direction such that Q_e equals Q_{min}, and equation (3-11) yields

$$p = 1.086\, N_g \sigma_g (Q_{max} - Q_{min}) \tag{3-13}$$

where equation (3-1) has been used to relate $\mathscr{F}_{min}$ to Q_{max} and $\mathscr{F}_{max}$ to Q_{min}, respectively.

In Section 4.5 it is shown that if the grains are elongated particles, which we may idealize as prolate spheroids, a strong magnetic field, **B**, can align these grains so that their major axes tend to be uniformly distributed in a plane perpendicular to the magnetic field. If the magnetic field is perpendicular to the line of sight, the plane of vibration of the polarized light will then be parallel to **B**, since for **E** parallel to the orienting magnetic field the electric vector tends to be perpendicular to the major axes of the grains, and the extinction is less than for **E** perpendicular to **B**.

The assumption that elongated, aligned grains are responsible for the polarization requires that the observed ratio of polarization to color excess be consistent with the optical properties of the grains. If the solid particles are assumed to be elongated dielectric cylinders, with m between 1.25 and 1.33, the numerical values of r_p in Table 3-1 are applicable if the alignment is perfect. The maximum value of p/E_{B-V} in equation (3-12) may be divided by R_V, defined in equation (3-10), and compared directly with these theoretical values of r_p. If we set R_V equal to 3, a minimum value, the maximum observed value of p/A_V equals 0.063, about equal to the theoretical value of r_p for x about 5. Since the actual grains cannot be perfectly aligned, and will be somewhat less effective polarizers than the ideal infinite cylinders, r_p will be less than the theoretical values in Table 3.1, and the effective value of x for the grains must be appreciably less than 5. Aligned cylinders with an index of refraction corresponding roughly to that of ice and with x equal to about 2 at 5500 Å, the wavelength for visual extinction measures, can apparently explain the observed value of p/E_{B-V} with a margin of safety of about 6. The corresponding value of a, the cylinder radius, is about 2×10^{-5} cm. Allowance for the somewhat smaller polarizing efficiency of spheroids with b/a equal to 2, for example, and for some deviation from perfect alignment indicates that such dielectric spheroids with this same value of a could be consistent with the observations. Evidently this argument yields only a rough upper limit on a. However, the variation of p with wavelength, discussed below, provides confirmation of this dimension for the spheroids that can be assumed.

3. OBSERVATIONS OF THE GRAINS

Dependence on Wavelength

Over the wavelength range from 3000 to 8000 Å the polarization is observed to be remarkably flat, falling off at greater λ. The mean curve is shown in Figure 3.4, where the different symbols represent

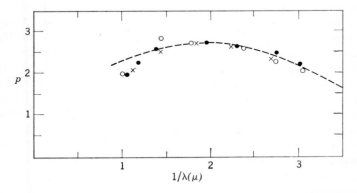

Fig. 3.4. Dependence of polarization on wavelength. The mean polarization on an arbitrary scale is plotted against the reciprocal wavelength in microns; different symbols represent results by different observers, all taken from ref. [2]. The dashed curve represents $Q_{eE} - Q_{eH}$ from Fig. 3.2, with a cylinder radius of 0.2 μ, but has been smoothed to take into account some dispersion in particle sizes.

independent sets of data [2]. Evidently p is an entirely different function of wavelength from the interstellar extinction, A_v, shown in Figure 3.3. This behavior would seem to exclude immediately polarization by refracting particles of radius small compared to the wavelength. For such Rayleigh-Gans particles, characterized by x small compared to unity, both p and A_v should increase sharply with decreasing wavelength. Similarly, small graphite particles can probably be excluded as the chief polarizing agents.

The wavelength variation in Figure 3.4 is roughly consistent with the values of $Q_{eE} - Q_{eH}$ shown in Figure 3.2 for dielectric cylinders with a refractive index of 1.33. The dashed curve in Figure 3.4 has been transferred from Figure 3.2 with an assumed cylinder radius of 2.0×10^{-5} cm, somewhat less than the spherical radius of 2.9×10^{-5} cm which analysis of the selective extinction data indicated for the dominant constituent of the grains. In fact a distribution of particle sizes must be present. Since the total extinction increases much more rapidly with particle radius than does the polarization, it is not surprising that the mean radius for extinction should exceed that for polarization. More detailed computations for cylinders of ice [2], taking into account the variation of the refractive index with wavelength and a distribution of cylindrical radii, confirm that rough agreement between theory and observation can be obtained for both extinction and polarization, provided that the cylinders are reasonably well aligned. There seems no reason to doubt that the observed interstellar extinction and polarization could be produced by elongated dielectric grains, though significant contributions by more complex core-mantle grains or by free radicals cannot be excluded.

Dependence on Galactic Longitude

The observed polarization of a group of stars has markedly different statistical properties for different directions in the sky. To analyze these effects it is customary to introduce the Stokes parameters, q and u, in magnitude units, defined by the relations

$$q = p \cos 2(\theta_p - \pi/2) \tag{3-14}$$

$$u = p \sin 2(\theta_p - \pi/2) \tag{3-15}$$

where θ_p is the position angle of the plane of vibration of the polarized radiation, measured here in galactic coordinates; thus θ_p is the angle between the plane of vibration projected on the plane of the sky and the great circle from the star to the north galactic pole. Evidently q is positive if this plane of vibration is parallel to the galactic plane ($\theta_p = \pi/2$), and negative if perpendicular to this plane ($\theta_p = 0$). When the line of sight to a star passes through several clouds, each producing a polarization p_j with a position angle θ_{pj}, it may be shown [25] that the resulting values of q and u

3. OBSERVATIONS OF THE GRAINS

are simply the sums of q_j and u_j over all clouds, provided the polarization is small, as is generally the case.

Mean values of q as a function of galactic longitude, l, for stars further away than 600 pc and with b less than 3°, are shown in Figure 3.5 [23], where each point represents an average over some 25 stars. The double-sine-wave curve drawn in the figure corresponds to what would be expected if the grains were prolate spheroids, oriented with their major axes perpendicular to a magnetic field in the direction of l equal to 50°; evidently the mean q falls to a relatively low value between the maxima. For the closer stars the observed minima in q are more nearly at about 60° and 240°. The direction of the Orion arm is roughly 70°, with values differing by some $\pm 15°$ for different determinations. The magnetic

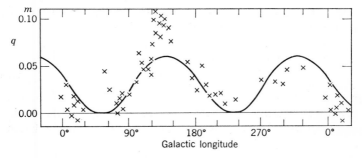

Fig. 3.5. Dependence of polarization on galactic longitude. Mean values of the Stokes parameter q (the component of p parallel to the galactic plane) are plotted for stars in different intervals of galactic longitude, taken from ref. [23]. Each point represents an average for a group of 20-25 stars; all stars are at galactic latitudes less than 3° and at distances more than 500 pc. The double sine wave is arbitrarily drawn with minima at $l = 50°$ and 230°.

field directions determined from these optical polarization data, from the Faraday rotation measurements in external galaxies, and from the polarization of galactic synchrotron emission (see Section 2.4) are somewhat discordant, with a field indicated in the direction of l equal to 50°, 100° and 70° from these three sets of data, respectively. While the observational uncertainty is appreciable, the magnetic field configuration is certainly more complicated than the simple uniform **B** assumed in analyzing the data.

The optical polarization observations give quantitative data on the nonuniformities in the magnetic field. The mean value of u is nearly zero for stars at low galactic latitude, as would be expected for a mean magnetic field parallel to the galactic plane. However, the r.m.s. dispersion of u, denoted by σ_u, is appreciable as a result of the different directions of **B** in different regions. The magnitude of σ_u may be used to determine the r.m.s. fluctuations of the position angle θ_{pj} from one region to another, provided a number of statistical assumptions are made. An idealized model will be assumed in which the polarization is produced by a random distribution of clouds, each producing the same p_j, but with a distribution of θ_{pj} about some mean value. We consider a group of stars in the galactic plane, all at the same distance, viewed in a direction perpendicular to **B**. If **B** is parallel to the galactic plane, the mean θ_{pj} is $\pi/2$ radians, and we define

$$\alpha^2 = \langle (\theta_{pj} - \pi/2)^2 \rangle \tag{3-16}$$

where the brackets denote a mean value for all the stars considered. Evidently, α is the dispersion of θ_{pj} in radians. On the assumption that the values of θ_{pj} in different clouds along the line of sight are uncorrelated, equation (3-15) yields

$$\sigma_u^2 = p_j^2 \sum_j \langle \sin^2 2(\theta_{pj} - \pi/2) \rangle = \frac{4p^2\alpha^2}{n} \tag{3-17}$$

where n is the average number of clouds along the line of sight, equal to p/p_j; to obtain equation (3-17), use has been made of equation (3-16), together with the assumption that $\theta_{pj} - \pi/2$ is small compared to unity.

To determine n we use the dispersion of the q values, denoted by σ_q. For $\theta_{pj} - \pi/2$ small the cosine in equation (3-14) is unity, and q equals the sum of the p_j. Since the r.m.s. dispersion in the number of clouds along the line of sight is $n^{1/2}$, we have

$$\sigma_q^2 = np_j^2 = \frac{p^2}{n} \tag{3-18}$$

3. OBSERVATIONS OF THE GRAINS

Equations (3-17) and (3-18) now yield

$$\frac{\sigma_u}{\sigma_q} = 2\alpha \qquad (3\text{-}19)$$

The dispersions σ_u and σ_q have been determined for the stars in the double cluster h and χ Per, with an observed ratio of 1.0 for σ_u/σ_q, giving a value of about 0.5 for α, or roughly 30°. The above analysis is rather approximate for such large α, but a more detailed analysis [26] yields about the same result. The corresponding ratio of the fluctuating magnetic field in one direction to the mean field is about 0.6. For stars more than about a parsec apart, the fluctuations of q and u show little correlation, indicating that the scale size for these fluctuations in magnetic field is less than a parsec. The value of n found from equation (3-18) is about 100 for this double cluster, consistent with a small scale for the magnetic field fluctuations—about an order of magnitude smaller than for the fluctuations in selective extinction (see below).

These numerical values depend, of course, on the statistical model used. Other assumptions would give somewhat different results. In particular, if the number n of statistically independent regions is set equal to 10 in equation (3-17), a value more typical for the interstellar medium generally, α is reduced to about 0.2 for the double cluster in Perseus, corresponding to about 10°. Similar values are obtained in this way from other regions [23], and it seems very likely that the ratio of the fluctuating magnetic field in one direction to the mean field is at least 0.2 and may be as great as 0.6. As pointed out in Section 2.4, fluctuations over a relatively small scale are indicated also by variations in the angle of polarization of the galactic radio noise, although quantitative data are lacking. Quite probably fluctuations of **B** are present with a wide variety of scales.

3.4 Spatial Distribution of Interstellar Matter

The observational data discussed in Section 2.2 indicate that the density of neutral hydrogen in the galactic plane, averaged over a ring some 10 kpc from the galactic center is about 0.8×10^{-24}

g/cm^3. Assumption of a helium–hydrogen ratio, Y/X, of 0.40 by mass (see Table 4.7) increases the gas density to about 1.1×10^{-24} g/cm^3. Within 1000 pc of the Sun, including part of the Orion spiral arm, the hydrogen density is somewhat greater and in Section 5.1 a mean gas density of 2×10^{-24} g/cm^3 is adopted as standard. As shown in Section 3.2, measures of E_{B-V} indicate that the density of the solid particles, which we shall here take to be dielectric spheroids, is about 1.3×10^{-26} g/cm^3; with some allowance for particles of larger radius suggested by extinction measures in the infrared, the density of dust may be taken as roughly 1 % of the adopted value for the gas density. The nonuniform distribution of this material has already been noted. In external galaxies the regions of $H\alpha$ emission and of extinction by grains are strongly concentrated in spiral arms, and the 21-cm data in our own Galaxy are consistent with a similar concentration of the neutral gas to spiral arms.

In the present Section evidence is presented on the nonuniform distribution of gas and dust in the solar neighborhood. Spiral structure is not discussed here, but irregularities of smaller scale are considered. Information on these nonuniformities is obtained both directly from observations of the bright and dark nebulae seen in photographs of the Milky Way and indirectly from analyses of irregularities found in most of the data concerned with interstellar matter, including extinction, line absorption and line emission.

Visible Nebulae

Even a casual glance at Milky Way photographs [27] shows that bright and dark features with a wide variety of sizes and shapes can be discerned. Rough estimates of distance to these features confirm that structures of all sizes are present in interstellar space, from vast extended obscuring regions many parsecs in size down to emission filaments less than 0.01 pc wide. Typical features around an O-type star, including "elephant-trunk" structures, are shown in the frontispiece.

3. OBSERVATIONS OF THE GRAINS

A rough classification of the most prominent types of dark nebulae is given in Table 3.2 [28]. The masses are obtained by

TABLE 3.2

Visible Dark Nebulae

	Small globule	Large globule	Intermediate cloud	Large cloud
$M/M_\odot$	>0.1	$3:$	8×10^2	1.8×10^4
R(pc)	0.03	0.25	4	20
n(H/cm^3)	$>4 \times 10^4$	$1.6 \times 10^3:$	100	20
$\dfrac{M}{\pi R^2} \dfrac{\text{g}}{\text{cm}^2}$	$>10^{-2}$	$3 \times 10^{-3}:$	3×10^{-3}	3×10^{-3}

estimating the mass of the grains required to explain the observed extinction of about 1.5 magnitudes in the three larger nebulae, and then multiplying by 100, the rough estimate obtained above for the ratio of gas to grains. A more detailed study [17] of dark nebulae shows a very large variety of structures, with smaller nebulae or clouds showing some tendency to cluster together in vast cloud banks or cloud complexes, through which the total extinction may be some 3 mag or more.

Bright nebulae are somewhat more difficult to interpret in terms of the density distribution, since the observed intensity depends on the presence of a bright star nearby as well as on the presence of gas and dust. However, observed emission nebulae seem to correspond roughly with the dark objects listed in Table 3.2. For example, the radius and density of the Orion nebula are similar to those of the large globules, while some of the extended H II regions have radii and densities comparable with those of the large cloud.

The occurrence of narrow elongated streamers or filaments is a marked characteristic of certain nebulae, including, for example, the Pleiades. For stars seen through such filamentary nebulae there tends to be a good agreement between the orientation of the plane of vibration and the observed direction of a filament [23]; i.e., the electric vector **E** in the radiation tends to be parallel to the filament. Thus if the grains are prolate spheroids oriented with their major axes perpendicular to a magnetic field, **B**, then **B** tends to be parallel to the filaments.

Indirect and Statistical Evidence

The irregular distribution of interstellar matter is indicated both by the velocity profiles of absorption lines and by the distribution of color excesses observed for stars at about the same distance. As noted in Section 2.3, separate components are observed in the absorption lines of interstellar Ca II and in the 21-cm line of neutral H. One may conclude definitely that the velocity pattern of the interstellar gas is irregular, with different regions characterized by quite different values of the fluid velocity, as measured with respect to the local standard of rest. The number of such regions along a line of sight in the galactic plane is about 8 per kpc, and the dispersion of radial velocities between different regions is about 8 km/sec.

Measures of selective extinction give information on the granularity of the dust distribution. If the dust were uniformly distributed, the dispersion of color excesses for stars at a fixed distance would be very small. In fact this dispersion is large, and its value may be used to give some information on the clumpiness of the dust distribution. To make comparison with a definite if hypothetical model, one may assume that there are separate clouds distributed at random, that a line of sight in the galactic plane intersects k such clouds per kpc, on the average, and that each such cloud produces a color excess E_0 on the B-V system. The probability $p(n)$ that a line of sight along a distance r will intersect n clouds is given by the usual Poisson distribution

$$p(n) = \frac{(kr)^n e^{-kr}}{n!} \qquad (3\text{-}20)$$

The average color excess at the distance r is given by

$$\langle E_{B\text{-}V} \rangle = krE_0 \qquad (3\text{-}21)$$

3. OBSERVATIONS OF THE GRAINS

For the average value of E_{B-V}^2 we obtain

$$\langle E_{B-V}^2 \rangle = \sum_{n=1}^{\infty} (nE_0)^2 p(n) \tag{3-22}$$

Substitution from equation (3-20) and evaluation of the sum yield

$$\langle E_{B-V}^2 \rangle = \{(kr)^2 + kr\} E_0^2 \tag{3-23}$$

from which one obtains the result

$$E_0 = \frac{\langle E_{B-V}^2 \rangle}{\langle E_{B-V} \rangle} - \langle E_{B-V} \rangle \tag{3-24}$$

where $\langle \ \rangle$ denotes an average taken over all stars at the same distance. Once E_0 is known, k may be obtained from equation (3-21).

Equation (3-24) has been used [8] to compute E_0 and k for some 500 stars within 1000 pc of the Sun and within 100 pc of the galactic plane, giving 0.14 mag for E_0 and 4.3 for k. The mean coefficient of selective extinction, equal to $E_0 k$, has already been used in equation (3-9). The observed distribution of E_{B-V} has a greater skewness than would be expected for clouds all of one type. The observed value of $\langle E_{B-V}^3 \rangle$ can be fitted if two types of clouds are assumed; the resultant values of E_0 and k for each of the two cloud types, here called "standard clouds" and "large clouds," are given in Table 3.3. The errors given are average deviations for the two distance groups analyzed. A correction for the dispersion of distance within each groups of stars was applied in computing the values in Table 3.3, which are somewhat uncertain. Confirmatory evidence that this statistical picture is realistic is provided by data in directions where no extended absorbing cloud complexes seem to be present, where analysis in terms of a single type cloud gives roughly the parameters for the standard cloud in Table 3.3. A similar analysis [29] which takes into account an assumed spherical shape for the clouds gives about the same results as in Table 3.3.

The total extinction in each cloud has been determined in a similar manner from the dispersion of galactic counts [30] and from the dispersion of star counts in the galactic plane [31]. The

TABLE 3.3

Statistical Properties of Dust Clouds

Type of cloud	Standard cloud	Large cloud
Mean E_{B-V} per cloud, E_0	0.061 ± 0.006	0.29 ± 0.06
Number of clouds per kpc, k	6.2 ± 0.3	0.8 ± 0.2
Selective extinction per kpc, kE_0	0.38 ± 0.05	0.23 ± 0.01

results are generally consistent with a total photographic extinction of 0.2 mag per cloud, but observational uncertainties and unknown intrinsic dispersions in the number of galaxies and stars per unit solid angle make these results less reliable than those given in Table 3.3.

The large clouds indicated in Table 3.3 would each have a total extinction of about 1 mag, roughly the same as for the three larger cloud types in Table 3.2. The spatial correlation of color excesses shows [32] strong reddening over areas several degrees across and gives a mean radius of about 35 pc for these "large clouds" in Table 3.3. Evidently these structures are somewhat similar to the "large cloud" in Table 3.2.

The standard clouds, each with a visual extinction of only about 0.2 magnitude, cannot be observed directly, but the color excess data seem to indicate fairly clearly that separate structures of about this opacity must exist between the larger clouds or cloud complexes. As a working hypothesis it seems reasonable to identify the standard clouds with the separate regions producing absorption lines in Ca II and neutral H, since the number k per pc is about the same, and since the observational evidence suggests some correlation between grains and gas. While k and A_c have been evaluated directly, the mean radius R of such clouds, which is related both to the density within the cloud and the fraction F_c of volume in the galactic plane occupied by such clouds, has not been determined with any precision. The spatial correlation of color excesses does not give very trustworthy results for these standard clouds,

3. OBSERVATIONS OF THE GRAINS

since the mean radius obtained [32] varies from about 5 pc for stars at some 1000 pc to less than 1.5 pc for relatively close stars. If diffuse nebulae are assumed to be standard clouds that happen to be close enough to a bright star to become luminous, then the fraction of these clouds that should be visible as reflection or emission nebulae has been estimated [33] as 5×10^{-4}; the observed number of diffuse nebulae then gives 1.2×10^5 (kpc)$^{-3}$ for n_s, the number of standard clouds per unit volume. This number may be too high, since it neglects the tendency for young stars to be concentrated in regions containing interstellar gas, and the corresponding values of 4.6 pc for R and 0.05 for F_c seem rather low. Alternatively the analysis of ionization equilibrium of Na gives a higher value for F_c in equation (4-66), if minimum ionization by cosmic rays is assumed. Neither of these values for F_c is reliable, and we shall here take an arbitrary mean of 0.07 for F_c to indicate in Table 3.4 the values of the other parameters which follow from

TABLE 3.4

Parameters of a Standard Cloud

Radius, R	7 pc
No. of clouds per (kpc)^3, n_s	5×10^4
No. in line of sight per kpc, $k = \pi R^2 n_s$	8
Fraction of volume occupied, $F_c = 4\pi R^3 n_s/3$	0.07
Visual extinction in a single cloud, A_c	0.2 mag
Density of H, n_{Hc}	10 cm^{-3}
Density of heavy ions, n_{ic}	5×10^{-3} cm^{-3}
Mass, M_c	400 $M_\odot$

this choice. The mean value of n_H resulting from standard clouds in the galactic plane, equal to $F_c n_{Hc}$, has been set equal to 0.7 cm^{-3}. This value exceeds the average density of 0.5 cm^{-3} found for the galactic disc generally (see Section 2.2) allowing for some concentration is spiral arms. The value may underestimate the local H density, but not all H atoms will be found in standard clouds. The density of positive ions heavier than hydrogen and helium follows from n_{Hc} and from the chemical composition adopted in the following chapter (see Table 4.7). The electron density, n_{ec}, may substantially exceed n_{ic}, if ionization of H by energetic particles is appreciable.

The values of F_c, R, and n_{Hc} in Table 3.4 should be regarded as illustrative only. While k and A_c are relatively well determined, the other interrelated quantities are not really known. In any case the parameters of the standard cloud must represent averages taken over a continuous range of values; the appropriate average will be somewhat different for each type of observation.

Between the clouds the density of gas and dust will not, of course, be zero. It was pointed out in Section 2.3 that the mean density of neutral sodium and ionized calcium along the line of sight to several adjacent stars is a full order of magnitude below its mean value. The analysis of ionization equilibrium given in Section 4.3 indicates that n_H is such regions cannot exceed 0.03 cm^{-3}, though uncertainties in the chemical composition make even this upper limit somewhat unreliable. Theoretical arguments on pressure equilibrium, presented in Section 5.2, suggest that n_H between the clouds is a few percent of its value within the clouds, or roughly 0.3 cm^{-3}. As noted in Section 2.2, observations of 21-cm emission from intermediate galactic latitudes indicate that a relatively uniform layer of neutral H, with $n(H\ I)$ in the range between 0.1 and 0.3 cm^{-3}, may extend several hundred parsecs above the galactic plane, and that concentrations of gas, or clouds, may be relatively less important at these heights than in the galactic plane. Evidently the average value of n_H between the clouds is quite uncertain, but probably lies within the range from 0.03 to 0.3 cm^{-3}.

References

1. H. C. van de Hulst, *Light Scattering by Small Particles*, Wiley, New York, 1957.
2. J. M. Greenberg, in *Stars and Stellar Systems*, Vol. 7, Univ. of Chicago Press, Chicago, 1968, p. 221.
3. R. Cayrel and E. Schatzman, *Ann. Astrophys.*, **17**, 555 (1954).

3. OBSERVATIONS OF THE GRAINS

4. N. C. Wickramsinghe, *Interstellar Grains*, Chapman and Hall, London, 1967.
5. J. R. Platt, *Astrophys. J.*, **123**, 486 (1956).
6. S. Sharpless, in *Stars and Stellar Systems*, Vol. 3, Univ. of Chicago Press, Chicago, 1963, p. 225.
7. J. Stebbins, C. M. Huffer and A. E. Whitford, *Astrophys. J.*, **91**, 20 (1940). **92**, 193 (1940).
8. G. Münch, *Astrophys. J.*, **116**, 575 (1952).
9. H. L. Johnson, in *Stars and Stellar Systems*, Vol. 3, Univ. of Chicago Press, Chicago, 1963, p. 204.
10. H. Arp, *Astrophys. J.*, **135**, 971 (1962).
11. H. L. Johnson, in *Stars and Stellar Systems*, Vol. 7, Univ. of Chicago Press, Chicago, 1968, p. 167.
12. A. E. Whitford, *Astron. J.*, **63**, 201 (1958).
13. H. C. van de Hulst, *Rech. Astron. Obs. Utrecht*, **11**, Part 2 (1949).
14. A. Güttler, *Z. Astrophys.*, **31**, 1 (1952).
15. T. P. Stecher, *Astrophys. J.*, **142**, 1683 (1965).
16. H. Neckel, *Z. Astrophys.*, **62**, 180 (1965).
17. B. T. Lynds, in *Stars and Stellar Systems*, Vol. 7, Univ. of Chicago Press, Chicago, 1968, p. 119
18. L. G. Henyey and J. L. Greenstein, *Astrophys. J.*, **93**, 70 (1941).
19. C. R. O'Dell, *Interstellar Grains*, (Symposium at Troy, N.Y., 1965; U.S. Gov't. Printing Office, NASA SP-140, 1967, p. 137.
20. H. M. Johnson, in *Stars and Stellar Systems*, Vol. 7, Univ. of Chicago Press, Chicago, 1967, p. 65
21. W. A. Hiltner, *Science*, **109**, 165 (1949).
22. J. S. Hall, *Science*, **109**, 166 (1949).
23. J. S. Hall and K. Serkowski, in *Stars and Stellar Systems*, Vol. 3, Univ. of Chicago Press, Chicago, 1963, p. 293.
24. W. A. Hiltner, *Astrophys. J. Supp.*, **2**, 389 (1956), No. 24.
25. S. Chandrasekhar, *Radiative Transfer*, Clarendon Press, Oxford, 1950, Section 15.
26. K. Serkowski, *Advan. Astron. Astrophys.*, **1**, 289 (1962).
27. E. E. Barnard, *A Photographic Atlas of Selected Regions of the Milky Way*, Carnegie Institution of Washington, 1927.
28. L. Spitzer, in *Stars and Stellar Systems*, Vol. 7, Univ. of Chicago Press, Chicago, 1968, p. 1.
29. E. Schatzman, *Ann. d'Astrophys.*, **13**, 367 (1950).
30. V. A. Ambartsumian, *Bull. Abastumani*, **4**, 17 (1940).
31. S. Chandrasekhar and G. Münch, *Astrophys. J.*, **112**, 380 (1950); **115**, 103 (1952).
32. H. Scheffler, *Z. Astrophys.*, **65**, 60 (1967).
33. V. A. Ambartsumian and S. G. Gordeladse, *Bull. Abastumani*, **2**, 37 (1938).

Chapter 4

Interactions among Interstellar Particles

4.1 Collisional Processes

Since the interstellar gas is far from thermodynamic equilibrium in many respects, an understanding of this environment requires a detailed analysis of the individual physical processes going on. Because of the very low density these processes mostly involve two-body collisions among photons, electrons, atoms, molecules and grains. Interactions between photons on the one hand and atoms, molecules, and grains on the other have been discussed in order to understand the observational data. In this chapter the interstellar role played by other types of collisions is analyzed, including particularly their effects on the local equilibrium of interstellar matter, excluding large-scale dynamical phenomena. As an introduction to this discussion the present section summarizes physical information on some of the the most important types of collisions.

Following a standard terminology [1] we shall call a "test particle" a particle which is being followed in space and time, and whose velocity and excitation are being changed by encounters with "field particles." The random velocities of the test and field particles will be denoted by $\mathbf{w}$ and $\mathbf{w}_f$, respectively, with their relative velocity $\mathbf{w} - \mathbf{w}_f$ denoted by $\mathbf{u}$. If n_f denotes the field particle density, and if we assume for the moment that $\mathbf{u}$ is the same for

4. INTERACTIONS AMONG INTERSTELLAR PARTICLES

all field particles, then the probability that the test particle experiences a collision in a time dt equals $C(u)dt$, where

$$C(u) = n_f u \sigma(u) \tag{4-1}$$

$\sigma(u)$ is the "collision cross section." The test particle may be regarded as located at the center of a circle of area σ, whose plane is perpendicular to $\mathbf{u}$. The cylindrical volume swept over in a time dt is $\sigma u dt$, and the number of field particles within the volume, equal to the average number of collisions within the time dt, is then $C(u)dt$, with C given in equation (4-1).

The probability per unit time of atomic or molecular excitation from level j to level k as a result of impact with some type of field particle is expressed as $n_f \gamma_{jk}$, where γ_{jk} is the "rate coefficient" for the excitation process; this probability is computed by averaging $C_{jk}(u)$ over all relative velocities, where $C_{jk}(u)$ denotes the value of $C(u)$ for excitation from level j to level k. If the velocity distributions are Maxwellian at some temperature, T, then the fraction of all encounters whose velocity is in the volume element du will be given by the usual Maxwellian distribution function

$$f(u)\,du = \frac{l^3}{\pi^{3/2}} e^{-l^2 u^2}\,du_x\,du_y\,dy_z \tag{4-2}$$

where l is defined by

$$l^2 \equiv \frac{m_r}{2kT} = \frac{m m_f}{2(m+m_f)kT} \tag{4-3}$$

with m_r representing the reduced mass, and with m and m_f representing the mass of the test and field particles, respectively. If we multiply $C_{jk}(u)/n_f$ by $f(u)du$, and integrate over all du, assuming that the excitation cross section, $\sigma_{jk}(\mathbf{u})$, is independent of the direction of $\mathbf{u}$, we obtain

$$\gamma_{jk} \equiv \langle \sigma_{jk} u \rangle = \frac{4}{\pi^{1/2}} l^3 \int_0^\infty u^3 \sigma_{jk}(u) e^{-l^2 u^2}\,du \tag{4-4}$$

where $\langle\ \rangle$ denotes an average over a Maxwellian distribution.

It is sometimes desirable to know the mean $C(u)$ for test particles of a particular velocity w, averaged over a Maxwellian velocity distribution for the field particle. If we multiply $C(u)$ by $f(w_f)dw_f$, and integrate over dw_f we obtain a mean $C(w)$; the functional form of $f(w_f)$ is given in equation (4-2), with w_f and m_f replacing u and m, in equations (4-2) and (4-3). The mean time, $t(w)$, between collisions, a function of the velocity w of the test particles, is given by

$$t(w) = \frac{1}{C(w)} \qquad (4\text{-}5)$$

Values of cross sections for some of the more important interstellar processes and the resultant values of $t(w)$ and γ_{jk} are discussed below.

Elastic Collisions, Short-Range Forces

When the kinetic energy of two colliding particles is the same after the collision as before, the collision is said to be elastic, though energy is generally transferred from one particle to another. We consider first the situation in which one of the particles is electrically neutral, and the forces are relatively short range, postponing until later the discussion of interactions between electrically charged particles, characterized by long-range, inverse-square forces.

Where the forces are relatively short range, the colliding particles behave somewhat as rigid spheres. We shall give approximate equations here for some of the times that appear in the theory of colliding spheres, following these with numerical values of the cross sections appropriate for collisions involving neutral hydrogen, the only species of neutral atom considered in detail here.

When particles are interacting with each other in a perfect gas, collisions will tend to establish a Maxwellian velocity distribution. After the particles have been deflected 90°, on the average, in mutual collisions one would expect any deviations from a Maxwellian velocity distribution to be much reduced. As a measure of

4. INTERACTIONS AMONG INTERSTELLAR PRATICLES

this tendency the "self-collision" time, t_c, is defined as the time in which particles of one type and a kinetic energy equal to the mean value experience a deflection of 90° or more in a single collision with another particle of the same type. From equations (4-1) and (4-5) it follows that the value of t_c is related to σ_c, the corresponding cross section for a deflection of at least 90°, by the relation

$$t_c = \frac{1}{2^{1/2} w_m n_f \sigma_c} \tag{4-6}$$

where w_m is the r.m.s. particle velocity, and in accordance with equations (4-2) and (4-3) $2^{1/2} w_m$ is the r.m.s. relative velocity between the particles.

For some purposes we are interested not in the randomization of velocities but rather in the average exchange of momentum between heavy test particles and a group of less massive field particles. This quantity is needed for computing the drift velocity either of atomic ions or of solid particles in a gas of neutral hydrogen. If $\Delta \mathbf{w}$ is the average change of velocity of a test particle in a single encounter, then we define a "slowing-down" time, t_s, so that w/t_s equals the average change of velocity of a test particle per unit time resulting from collisions, and write

$$\frac{1}{t_s} = -\frac{1}{w} n_f |\langle \sigma u \, \Delta \mathbf{w} \rangle| = -\frac{1}{w} \left| \frac{d\mathbf{w}}{dt} \right| \tag{4-7}$$

where the brackets again denote an average over a Maxwell-Boltzmann distribution of u, as in equation (4-4). Since the field particles are assumed to be much lighter than the test particles, we may assume that w_f exceeds w and replace u and m_r by w_f and m_f, respectively. The mean velocity change $\Delta \mathbf{w}$ may be written approximately

$$\Delta \mathbf{w} = -\frac{m_f \mathbf{w}}{m} \tag{4-8}$$

since the mean momentum transferred to a heavy test particle by a light field particle is about $-m_f \mathbf{w}$. A numerical constant of order unity, depending on the detailed nature of the mutual force between the two colliding particles, has been omitted from equation (4-8). Finally t_s becomes

$$\frac{1}{t_s} = \frac{n_f m_f}{m} \langle \sigma w_f \rangle = \frac{2 n_f \sigma_s m_f}{m} \left(\frac{2kT_f}{\pi m_f}\right)^{1/2} \quad (4\text{-}9)$$

In the second expression the elastic cross section, σ, has been taken out of the brackets and set equal to its average value, denoted by σ_s. The average deceleration of a group of test particles, all of the same initial velocity, **w**, is referred to as "dynamical friction." The slowing-down time is independent of the test particle's velocity, provided that this is much smaller than the mean velocity of the field particles.

The change of kinetic energy in a single encounter is related to $\Delta \mathbf{w}$ by the equation.

$$\Delta E = m\mathbf{w} \cdot \Delta \mathbf{w} + \tfrac{1}{2} m (\Delta w)^2 \quad (4\text{-}10)$$

If we now wish to compute dE/dt by summing equation (4-10) over all encounters per second, the first term on the right-hand side gives $-2E/t_s$ from equation (4-7). [Components of $\Delta \mathbf{w}$ perpendicular to **w** average out to zero in equation (4-7), since the field particles are assumed to have an isotropic velocity distribution.] The second term in equation (4-10) gives a positive contribution, which may be shown to be independent of w for w much less than w_f. If we average dE/dt over all test particles, and over all the field particles as well, assuming Maxwellian velocity distributions for both sets of particles, at temperatures T and T_f, respectively, then the second term depends only on T_f and must equal the first when T equals T_f. Since the mean energy of the test particles is proportional to T, we obtain

$$\frac{dT}{dt} = -\frac{2}{t_s}(T - T_f) \quad (4\text{-}11)$$

The time $t_s/2$ for the heavier particles is sometimes called the "equipartition time," since this measures the rate at which two groups of particles of different masses approach equipartition of kinetic energy with each other.

4. INTERACTIONS AMONG INTERSTELLAR PARTICLES

We turn now to the value of σ to be used in the expressions above. For collisions of atoms and molecules with solid particles with a radius, a, of many Ångstrom units, σ may be set equal to the geometrical cross section πa^2, denoted by σ_g, since this distance much exceeds the range of interatomic forces.

Values of σ for elastic collisions involving H atoms are given in Table 4.1. The first row gives the cross sections to be used in

TABLE 4.1

Elastic Collision Cross Sections for H Atoms
Units of 10^{-15} cm^2

Temperature T (°K)	10°	30°	100°	300°	1000°
σ_c for H–H	1.46	1.70	1.55	1.26	1.22
σ_s for i–H	46	28	15	8.4	4.6
σ_s for H–e	4.1	4.1	4.0	3.9	3.8

equation (4-6) for the self-collision time, t_c; these values, computed by quantum mechanics [2], give the cross sections for deflections of 90° or more in the relative orbit between two H atoms, whose initial relative velocity has the r.m.s. value corresponding to the temperature, T. Within a factor 2, about the same values are obtained from classical orbit theory [3]. The other two rows in the table give the appropriate values of σ_s to be used in equation (4-9) to give the correct slowing-down time, t_s, for the heavier particles. The value of σ_s for positive ions moving through H atoms was computed [4] from classical orbits, using the known $U(r)$, which here varies as r^{-4}; a quantum-mechanical calculation of σ_s for H–H$^+$ encounters [5] agrees with the classical value for T equal to 100°K, although equation (4-9) is not strictly applicable in this case, when field and test particles have the same mass. For both H–H and i–H encounters the collision radius $(\sigma_s/\pi)^{1/2}$ agrees within a factor 2 with the distance at which the potential energy equals the initial kinetic energy of relative motion and varies as $T^{-1/6}$ and $T^{-1/4}$, respectively, for these two cases.

Because of its lower polarizability, a He atom colliding with ions has a σ_s about half the corresponding value for H [4]. Values of σ_s for H atoms moving through electrons were taken from a quantum-mechanical calculation [6]; electron exchange reduces the cross section below the value for heavy ions.

Elastic Collisions, Electrostatic Forces

These same concepts may be applied also to encounters between charged particles. Because of the long-range character of electrostatic forces, the cumulative effect of many distant encounters, each producing a relatively small change, usually outweighs the effect of a few close encounters, which produce large deflections and velocity changes. Hence equations (4-6) and (4-9) for t_c and t_s must both be modified. If the mass and charge for the field and test particles are m_f, m, $Z_f e$ and Ze, respectively, then the slowing-down time, t_s, defined as in equation (4-9) for a heavy charged test particle interacting with lighter field particles, becomes [1, 7] on integrating over collisions at all distances,

$$t_s = \frac{3m(2\pi)^{1/2}(kT_f)^{3/2}}{8\pi m_f^{1/2} n_f Z^2 Z_f^2 e^4 \ln \Lambda} = \frac{11.7 A T_f^{3/2}}{A_f^{1/2} n_f Z^2 Z_f^2 \ln \Lambda} \text{ sec} \quad (4\text{-}12)$$

where A and A_f are the particle masses in atomic weight units. The quantity Λ, defined by

$$\Lambda = \frac{3}{2 Z Z_f e^3} \left(\frac{k^3 T^3}{\pi n_e} \right)^{1/2} \quad (4\text{-}13)$$

is a "cutoff" factor, resulting from the ineffectiveness of encounters beyond some critical distance [7]. For interstellar conditions, with n_e equal to 1 and T between 10^2 and 10^4, $\ln \Lambda$ is about 20 [7]. The value of t_s computed from equation (4-12) is shorter than that found from collisions producing deflections of $90°$ or more by a factor about $(8 \ln \Lambda)^{-1}$.

Electrons in a fully ionized gas lose energy almost entirely in collisions with other electrons, and for such collisions the value of t_s obtained from equation (4-12) is not strictly applicable. For an

4. INTERACTIONS AMONG INTERSTELLAR PARTICLES

energetic electron, whose velocity, w, is appreciably greater than the r.m.s. thermal speed, the rate of energy loss is given by

$$\frac{dE}{dt} = -\frac{4\pi e^4 n_e \ln(\Lambda mw^2/3kT)}{mw} \equiv -\frac{E}{t_E} \qquad (4\text{-}14)$$

where t_E defined in this equation is an energy loss time. If E exceeds about 40 eV, the cutoff factor in equation (4-14) must be modified [7].

The self-collision time for ions of charge Ze and atomic weight A may be defined as the time for such ions to experience a cumulative r.m.s. deflection of $\pi/2$ in collisions with other ions of the same Z and A at a temperature T [1, 7]. If A_f is set equal to A, equation (4-12) gives t_c with an accuracy of about 3%. According to equation (4-11) the equipartition time for the kinetic temperature T, of heavy ions, to approach the temperature T_f of the lighter ions equals one-half the value of t_s obtained from equation (4-12). If electrons and protons are compared, these various times are related approximately by the ratios

$$\frac{t_c(e\text{-}e)}{t_c(p\text{-}p)} = \frac{t_c(p\text{-}p)}{t_s(p\text{-}e)} = \frac{A_e^{1/2}}{A_p^{1/2}} = \frac{1}{43} \qquad (4\text{-}15)$$

where the symbols in parentheses denote test and field particles, respectively.

Excitation by Electron Impact

Collisions of electrons with atoms can produce either excitation or deexcitation. In thermodynamic equilibrium the condition of detailed balancing requires that the rates of these two processes must be equal for corresponding velocity intervals, a condition which gives a relationship between σ_{jk} and σ_{kj}, the cross sections for excitation and deexcitation, respectively. The relative velocity, u, may be set equal to the electron velocity, which we denote by w. If w_j represents the electron velocity before excitation, and w_k the velocity after excitation, then the number of upwards collisions per cm^3 per sec produced by electrons in the velocity range between w_j and $w_j + dw_j$ must equal the number of deexcitations produced by electrons in the velocity range between w_k and $w_k + dw_k$; hence we have

$$n_j^* n_e f(w_j) w_j dw_j \sigma_{jk}(w_j) = n_k^* n_e f(w_k) w_k dw_k \sigma_{kj}(w_k) \quad (4\text{-}16)$$

If we substitute from equations (2-9) and (4-2) for n_j^*/n_k^* and $f(w)$ in thermodynamic equilibrium, we obtain the condition [8]

$$g_j w_j^2 \sigma_{jk}(w_j) = g_k w_k^2 \sigma_{kj}(w_k) \quad (4\text{-}17)$$

where the condition

$$\tfrac{1}{2} m w_k^2 = \tfrac{1}{2} m w_j^2 - E_{jk} \quad (4\text{-}18)$$

has been used, with E_{jk} representing $E_k - E_j$. Excitation can occur only if the initial kinetic energy is greater than the threshold value E_{jk}. Substitution of equation (4-17) into (4-4) gives

$$g_j \gamma_{jk} = g_k \gamma_{kj} e^{-E_{jk}/kT} \quad (4\text{-}19)$$

For excitation of neutral atoms by electrons the deexcitation cross section, σ_{kj}, is generally finite at the threshold, where w_k vanishes, and σ_{jk} approaches zero in accordance with equation (4-17). For collisions with ions, however, σ_{jk} remains finite at threshold as a result of the electrostatic acceleration of the electron, and hence σ_{kj} varies as $1/w_k^2$.

It is customary [8, 9] to express these cross sections in terms of a collision strength, $\Omega(j, k)$, defined by

$$\sigma_{jk}(w_j) = \frac{\pi}{g_j} \left(\frac{h}{2\pi m w_j} \right)^2 \Omega(j, k) \quad (4\text{-}20)$$

Equation (4-17) indicates that $\Omega(j, k)$ for excitation equals $\Omega(k, j)$ for deexcitation. The collision strength is of order unity for most transitions. Values of $\Omega(j, k)$ may be computed in two ways. For transitions with a finite electric dipole moment ("radiatively permitted" transitions) the electric field of the passing electron can be regarded as producing a radiative transition. The value of $\Omega(j, k)$ is then related to the oscillator strength f_{jk} by the relation [10].

4. INTERACTIONS AMONG INTERSTELLAR PARTICLES

$$\Omega(j, k) = \frac{8\pi b}{3^{1/2}} \frac{g_j f_{jk}}{E_{jk}/E_1} \quad (4\text{-}21)$$

where E_1 is the binding energy of the neutral H atom. The quantity b is a quantum-mechanical correction factor which for neutral atoms vanishes at threshold, but which for positive ions varies slowly with w_j and equals about 0.2 for electron kinetic energies between one and two times the excitation energy.

For transitions which have no electric dipole moment ("radiatively forbidden" transitions) f_{jk} is very small. Collisional excitation then occurs primarily by exchange of spins between the incident and the atomic electron. Values of $\Omega(j, k)$ have been computed for this process by quantum mechanics [8, 11, 12]. Values for excitation from the ground level up to all levels below 4 eV for the singly ionized atoms C^+, N^+, O^+, Ne^+, and Si^+, and also O^{++}, are given in Table 4.2; these transitions are the most important ones under interstellar conditions (see Section 4.4). In each case the sum of the transition probabilities from the upper level down to all the lower levels [12-14] is also given in Table 4.2.

TABLE 4.2

Collision Strengths for Excitation by Electrons

No. of p electrons	Ion	Terms Lower		Upper	E_{jk}(eV)	$\Omega(j,k)$	A_{kj}(sec^{-1})
1,5	C II	$^2P_{1/2}$	–	$^2P_{3/2}$	0.0079	1.28	2.4×10^{-6}
	Ne II	$^2P_{3/2}$	–	$^2P_{1/2}$	0.097	0.32	8.6×10^{-3}
	Si II	$^2P_{1/2}$	–	$^2P_{3/2}$	0.036	7.7	2.1×10^{-4}
2	N II	$^3P_0$	–	$^3P_1$	0.0061	0.40	2.1×10^{-6}
		$^3P_0$	–	$^3P_2$	0.0163	0.28	7.4×10^{-6}
		$^3P_1$	–	$^3P_2$	0.0102	1.13	7.4×10^{-6}
		$^3P_0$	–	$^1D_2$	1.90	0.34	4.0×10^{-3}
	O III	$^3P_0$	–	$^3P_1$	0.014	0.38	2.6×10^{-5}
		$^3P_0$	–	$^3P_2$	0.038	0.21	1.0×10^{-4}
		$^3P_1$	–	$^3P_2$	0.024	0.95	1.0×10^{-4}
		$^3P_0$	–	$^1D_2$	2.51	0.27	2.8×10^{-2}
3	O II	$^4S_{3/2}$	–	$^2D_{5/2}$	3.31	0.85	4.2×10^{-5}
		$^4S_{3/2}$	–	$^2D_{3/2}$	3.31	0.57	1.8×10^{-4}

For these transitions we may take $\Omega(j, k)$ to be independent of w, and a direct integration of equation (4-4), with use of equation (4-20), gives the rate coefficient for deexcitation.

$$\gamma_{kj} = \frac{h^2 \Omega(j, k)}{g_k (2\pi m)^{3/2} (kT)^{1/2}} = 8.63 \times 10^{-6} \frac{\Omega(j, k)}{g_k T^{1/2}} \frac{\text{cm}^3}{\text{sec}} \quad (4\text{-}22)$$

The corresponding rate coefficient for excitation is obtained on substituting this value for γ_{kj} into equation (4-19).

Excitation by Atoms

Among the various collisional processes involving heavy particles those involving neutral H atoms are the most important because of the preponderant cosmic abundance of hydrogen. Cross sections for excitation by neutral H atoms have been estimated theoretically for three different processes which play a major role in H I regions: (a) excitation of hyperfine levels in H atoms [15]; (b) excitation of low-lying, fine-structure levels in ionized or neutral atoms [16, 17]; (c) excitation of the $J = 2$ level in H_2 [18], with an excitation potential of 0.0439 eV [19]. In processes (a) and (b) the electron spins are changed by the collision, while in (c) the rotational angular momentum is changed. Average deexcitation cross sections for these three processes, multiplied by the ratios of statistical weights in the upper level, k, and the lower level, j, are given in Table 4.3 for different values of the kinetic temperature. As in the excitation of neutral atoms by electrons, the deexcitation cross sections for these types of collisions, with no inverse-square force between the colliding particles, change less rapidly with energy near threshold than do the excitation cross sections.

The numerical values for process (b) are those computed [16, 17] for the deexcitation of the levels in C^+ and Si^+ (see Table 4.2 for terms and excitation energies) for which g_k/g_j equals 2. These values are about five times the computed cross sections [17] for the deexcitation of the $J = 1$ level in O I, with an excitation energy of 0.0197 eV and with g_k/g_j equal to 3/5. For the $J = 0$ level in O I, with an energy of 0.0281 eV above the ground level

4. INTERACTIONS AMONG INTERSTELLAR PARTICLES

TABLE 4.3

Deexcitation Cross Sections for H Atoms
Units of 10^{-15} cm^2

Temperature T (°K)	10°	30°	100°	300°	1000°
$\frac{g_k}{g_j} \sigma_{kj}$ for H–H	2.3	2.8	2.3	1.5	1.1
$\frac{g_k}{g_j} \sigma_{kj}$ for H–ion	33	18	10	6.0	3.3
$\frac{g_k}{g_j} \sigma_{kj}$ for H–H$_2$	0.09	0.09	0.12	0.19	0.41

and with g_k/g_j equal to 1/5, the values of $g_k\sigma_{kj}/g_j$ are about a third of those for the $J = 1$ level. The ratio g_k/g_j equals 3 and 5 for processes (a) and (c), respectively.

The excitation and deexcitation cross sections are related by equation (4-17), just as for electron collisions. If we set σ_{kj} equal to an appropriate mean value in equation (4-4), equation (4-19) yields for γ_{jk}, the excitation rate coefficient,

$$\gamma_{jk} = \langle u \rangle \frac{g_k \sigma_{kj}}{g_j} 10^{-5040 E_{jk}(\text{eV})/T} \quad (4\text{-}23)$$

where E_{jk} (eV) denotes the excitation energy in eV; the mean relative velocity, $\langle u \rangle$, for H atoms colliding with particles of mass Am_H is given by

$$\langle u \rangle = 2 \left(\frac{2kT}{\pi m_r} \right)^{1/2} = 1.46 \times 10^4 T^{1/2} \left(1 + \frac{1}{A} \right)^{1/2} \text{ cm sec}^{-1} \quad (4\text{-}24)$$

As in equation (4-3) m_r equals $m_\text{H} A/(1 + A)$.

Recombination and Ionization

The cross section for capture by a bare nucleus of an electron into a level of total quantum number n is given by

$$\sigma_{cn} = A_r \frac{v_1}{v} \frac{hv_1}{\frac{1}{2}mw^2} \frac{g_{fn}}{n^3} \tag{4-25}$$

where the "recapture constant," A_r, is given by

$$A_r = \frac{2^4}{3^{3/2}} \frac{he^2}{m^2 c^3} = 2.11 \times 10^{-22} \text{ cm}^2 \tag{4-26}$$

The kinetic energy of the electron before capture is $\frac{1}{2}mw^2$, and hv is the energy of the emitted photon, equal to $\frac{1}{2}mw^2 + hv_1/n^2$. The Gaunt factor, g_{fn}, for these free-bound transitions has been computed theoretically [20], and depends on both w and n. For electron energies not exceeding $10E_1$, g_{fn} is about unity.

For the capture of electrons by atomic ions, with one or more bound electrons, equations (4-25) and (4-26) frequently provide a good approximation if the electrons are captured in excited levels, which tend to be hydrogenic. For electrons captured by such ions directly in the ground level the capture cross section may be less by 1 or 2 orders of magnitude than the value computed from equation (4-25). Separate computation of σ_{c1} is required for each atomic structure. It is customary to give results in terms of the cross section s_{v1} for absorption of continuous radiation by an electron in the ground level, since this is related to σ_{c1} through Kirchhoff's law, equation (2-5).

The recombination coefficient, α_{rj}, for electron capture in level j of an ion in stage of ionization $r+1$ is given by

$$\alpha_{rj} = \langle w\sigma_{cj} \rangle \tag{4-27}$$

while the total recombination coefficient, α_r, is defined by

$$\alpha_r = \sum_j \alpha_{rj} \tag{4-28}$$

In the following discussion we shall generally consider only direct radiative capture, with emission of a photon. However, in complex atoms and in molecules recombination may occur in other ways with a higher probability. For example, when bound electrons are already present, an ion can capture an electron in any of many highly excited levels, the energy going into excitation of

4. INTERACTIONS AMONG INTERSTELLAR PARTICLES

inner electrons; this process is called "dielectric recombination" [21]. For molecular ions, dissociative recombination, in which the electron is captured and the energy leads to dissociation of the molecule, can be the dominant process; the corresponding cross section can be comparable with the geometrical cross section [22], greater by many orders of magnitude than the cross section for radiative capture.

Ionization of H by an energetic particle of charge Z, with a velocity equal to β times the velocity of light, has a cross section, σ_{HR}, equal to [23]

$$\sigma_{HR} = \frac{2\pi e^4 Z^2}{mc^2 \beta^2 E_1} \times 0.285 \left\{ \ln \frac{2mc^2 \beta^2}{E_1(1-\beta^2)} + 3.04 - \beta^2 \right\}$$

$$= \frac{1.23 \times 10^{-20} Z^2}{\beta^2} \left\{ \log \frac{\beta^2}{1-\beta^2} + 6.20 - 0.43\beta^2 \right\} \quad (4\text{-}29)$$

As in equation (2-66) the subscript R denotes cosmic radiation and energetic particles generally. The quantity E_1 is the ionization energy of hydrogen, and m is the electron rest mass; ln and log denote logarithms to the base e and 10, respectively. For protons this equation is valid for energies exceeding about 0.3 MeV, corresponding to a value of 0.025 for β. The total energy lost by an energetic nucleus per primary ionizing collision, taking into account the excitation energy lost to the H atoms as well as the ionization and kinetic energy of the secondary electrons, is about 60 eV for a primary energy of 10^6 eV per nucleon, increasing to about 75 eV as the primary energy increases to 10^9 eV per nucleon. For protons in the low MeV range each secondary produces about 0.67 additional ion pairs in hydrogen, on the average, and the corresponding energy lost by the energetic particles per ion pair produced is about 36 eV [24].

4.2 Excitation and Kinetic Equilibrium

We turn now to a discussion of the population densities of various energy levels under interstellar conditions. As in Section

2.1 we let n_j be the number per cm^3 of particles in a level j characterized by g_j quantum states all of about the same energy E_j. In a steady state the number of transitions into level j from all levels k per unit time must equal the corresponding number out of level j. We consider first only bound levels and assume that transitions between levels j and k are caused by three processes: spontaneous radiative transitions from level j to lower levels k, induced radiative transitions back and forth between levels j and k, and transitions induced by electron collisions, which also can go in either direction. From the definition of the Einstein probability coefficients the total number of induced radiative transitions per cm^3 per sec from j to k is equal to $n_j B_{jk} U_v$, where U_v is the density of radiant energy per cm^3 per unit frequency interval, related to the specific intensity, I_v, by

$$U_v = \frac{1}{c} \int I_v \, d\omega \qquad (4\text{-}30)$$

If we introduce the collisional rate coefficient, γ_{jk}, defined in the previous section, we have

$$n_j \left\{ \sum_k B_{jk} U_v + \sum_{k<j} A_{jk} + n_e \sum_k \gamma_{jk} \right\} = \sum_k n_k B_{kj} U_v + \sum_{k>j} n_k A_{kj} + n_e \sum_k n_k \gamma_{kj} \qquad (4\text{-}31)$$

where we have suspended the previous convention that E_k always exceeds E_j. Only electron collisions are considered here; other types of collisions are readily added. If there are m separate levels, the set of equations (4-31) provides $m - 1$ linearly independent equations, which determine the $m - 1$ ratios of the population densities to each other. If the coefficients in this equation are all known, solution of this set of equations is, in principle, straightforward; such solutions have been obtained for the population of H and He atoms for various interstellar conditions. Comparison [25] of these detailed results with spectroscopic measures of planetary nebulae and the Orion nebula reveals significant

4. INTERACTIONS AMONG INTERSTELLAR PARTICLES

discrepancies, suggesting that either the rate coefficients are not entirely correct or that other processes, not considered in equation (4-31), may be important.

If U_ν is replaced by $4\pi B_\nu(T)/c$, its value in thermodynamic equilibrium, where the Planck intensity at the temperature T is defined in equation (2-4), then the set of equations (4-31) is satisfied identically, provided we use equations (2-22) and (4-19) relating the three Einstein coefficients and the two collisional rate coefficients, respectively. In fact the requirement that in thermodynamic equilibrium at arbitrary temperature and density the left-hand and right-hand sides of equation (4-31) must be equal can be used to compute all the coefficients on one side of the equation from those on the other. With obvious modifications equation (4-31) can also include transitions up to and down from the free state (see the subsequent Section).

Velocity Distribution Function

We now apply equation (4-31) to consider the deviations from a Maxwellian distribution of velocities in the interstellar gas. If only elastic collisions occur, the velocity distribution is necessarily Maxwellian, but if the high-energy electrons excite atoms in inelastic collisions, without any corresponding superelastic collisions, in which the electron gains energy by deexciting an atom or ion, then the high-energy tail of the Maxwellian distribution may be depleted.

To analyze this effect a very simple model will be considered in which the atoms have only one level that can be excited, with an energy E_2 above the ground level. In this model the electrons will be divided into two populations. Those of kinetic energy greater than E_2, corresponding to a velocity w_2, will be assumed to have a number density n_2; these electrons, which can suffer inelastic collisions with atoms, will be called the "high-speed electrons." The "low-speed electrons" will include all those of energy less than E_2, and for these all collisions are elastic; their number density is n_1. The energy E_2 is assumed to be appreciably greater than the mean energy of the low-speed electrons. In kinetic equilibrium

$$\frac{n_2^*}{n_1^*} = \frac{f_2}{f_1} e^{-E_2/kT} \tag{4-32}$$

where f_2 and f_1 are the partition functions for the two groups, defined as in equation (2-11), with the energy referred to E_2 for the high-speed electrons and to zero for the low-speed electrons. We assume n_2^*/n_1^* is much less than unity.

We consider now the transitions of electrons from group 2 to group 1. Collisions of the high-speed electrons with the slower ones will bring the average high-speed electron down into the low-speed group in about the energy-loss time, t_E, defined in equation (4-14). Since most of the high-speed electrons will have a kinetic energy not much greater than E_2, one may evaluate t_E at the velocity w_2, and the rate of downwards transitions, $n_2/t_E(w_2)$, will appear on one side of equation (4-31). A second term will be the probability of exciting an ion with particle density n_i in an inelastic collision; since again most of the high-speed electrons have a velocity about equal to w_2, this probability equals $n_i w_2 \sigma_{jk}(w_2)$ per electron per second.

Two corresponding terms will appear on the other side of equation (4-31) for this simple model. To maximize the deviations from thermodynamic equilibrium we ignore the term resulting from superelastic collisions with atoms, with the assumption that any excited state decays by radiation long before another electron collision; the conditions under which this assumption is valid are discussed below. There must also be a term corresponding to acceleration of low-speed electrons up to high-speed as a result of collisions with other electrons; this term we evaluate by the condition that when the cross section for inelastic encounters vanishes and all collisions are elastic, n_2/n_1 is given by its value in kinetic equilibrium [equation (4-32)]. Equation (4-31) then yields

$$n_2 \left\{ \frac{1}{t_E(w_2)} + n_i w_2 \sigma_{jk}(w_2) \right\} = n_1 \left\{ \frac{f_2 e^{-E_2/kT}}{f_1 t_E(w_2)} \right\} \tag{4-33}$$

4. INTERACTIONS AMONG INTERSTELLAR PARTICLES

which becomes

$$\frac{n_2}{n_1} = \frac{n_2{}^*}{n_1{}^*} \frac{1}{1 + n_i w_2 \sigma_{jk}(w_2) t_E} \tag{4-34}$$

The deviations from thermodynamic equilibrium depend on the ratio of two downwards transition probabilities for the free electrons; one, corresponding to inelastic collisions with atoms, is not compensated by upwards transitions, in contrast to the other, resulting from elastic collisions with other electrons.

We may compute numerical values for this ratio. For excitation of the $^1D_2$ level in O III, with an energy of 2.51 eV, Table 4.2, together with equation (4-20), gives a value of 1.3×10^{-16} cm^2 for σ_{jk}. For electrons with a kinetic energy of 2.51 eV the time t_E, defined in equation (4-14), has the value $2.6 \times 10^4/n_e$ sec. The ratio of downwards probabilities which appears in the denominator of equation (4-34) becomes

$$n_i w_2 \sigma(w_2) t_E = 3.2 \times 10^{-4} \frac{n_i}{n_e} \tag{4-35}$$

Evidently, in H II regions, when n_e exceeds n_i, the density of O III ions, by about 10^3, relative deviations from a Maxwellian distribution of velocities are less than 10^{-6}. Another process which might affect the velocity distribution in H II regions is the radiative capture of electrons by protons. However, from equations (4-25) and (4-26) we see that the time for electron captures in the ground state is about $6 \times 10^{12}/n_p$ sec for electrons of 1 eV energy, greater than t_E by some 10^8; hence this effect is also entirely negligible.

In H I regions the temperature is lower, and excitation of C II ions by electrons must be considered. While the cross section for excitation of C II by an electron with an energy of 0.0079 eV is much greater than that for excitation of O III at threshold, $t_E n_e$ is very much less at the lower temperature and the numerical constant in equation (4-35) is reduced by about a factor 1/130. In this case, however, n_i is about equal to n_e, and relative deviations from a Maxwellian velocity distribution are about 1 order of magnitude greater than in H II regions, but still less than 10^{-5}.

Evidently the electrons in the interstellar gas will normally have a velocity distribution very close to the Maxwellian. A more exact analysis [26] gives similar results.

For neutral H atoms the situation is not quite so conclusive. Either C II ions or H_2 molecules can be excited by H atoms. As a measure of the relative deviation from a Maxwellian distribution we may take $nw_H \sigma_{jk} t_c$; in comparison with the quantity $n_i w_2 \sigma_{jk}(w_2) t_E$ in equation (4-34), n, the particle density either of C II ions or of H_2 molecules, has replaced n_i, the O III ion density; the hydrogen thermal velocity, w_H, has replaced the electron velocity; and the self-collision time, t_c, for H atoms, obtained from equation (4-6), has replaced the energy loss time, t_E, for electrons. The cross section for H–H encounters relevant for t_c in H I regions is given in Table 4.1. The cross section σ_{jk} for excitation by H atoms depends on the energy; to obtain upper limits we take the values of $g_k \sigma_{kj}/g_j$ listed in Table 4.3 for T equal to 30°. In this way we find

$$n\sigma_{jk} w_H t_c \leq \begin{cases} 7.0 n(\text{C II})/n(\text{H I}) & \text{for C II excitation} \\ 3.4 \times 10^{-2} n(H_2)/n(\text{H I}) & \text{for } H_2 \text{ excitation} \end{cases} \quad (4\text{-}36)$$

Since $n(\text{C II})/n(\text{H I})$ is less than 1/2000 in H I regions, the relative deviation from a Maxwellian velocity distribution for H atoms will be less than 1%. If $n(H_2)$ is less than $n(\text{H I})/4$, then the relative number of H atoms more energetic than 0.037 eV, the threshold energy for H_2 excitation [19], will likewise be within 1% of its value in thermodynamic equilibrium. The conclusion that atoms and electrons in the interstellar gas have velocity distributions close to Maxwellian is a fundamental one in all interstellar matter studies. In addition, the kinetic temperature tends to be the same for different types of particles, as shown in Section 4.4.

Excitation

We now apply equation (4-31) to the idealized problem of an atom where only two levels need be considered, the ground level and the first excited level, with population densities n_1 and n_2.

4. INTERACTIONS AMONG INTERSTELLAR PARTICLES

We assume that the radiation field is dilute black body radiation, with an energy density given

$$U_\nu = \frac{4\pi W B_\nu(T)}{c} = \frac{8\pi W h\nu^3/c^3}{e^{h\nu/kT} - 1} \tag{4-37}$$

where W is called the "dilution factor." If in addition to equation (2-22) equation (4-19) is also used, with a kinetic temperature assumed to be identical with that in equation (4-37), equation equation (4-31) becomes

$$\frac{b_2}{b_1} = \frac{n_e \gamma_{21}/A_{21} + We^{h\nu/kT}/(e^{h\nu/kT} - 1)}{1 + n_e \gamma_{21}/A_{21} + W/(e^{h\nu/kT} - 1)} \tag{4-38}$$

where equations (2-9) and (2-15) have been used to express n_2/n_1 as their ratio in thermodynamic equilibrium multiplied by b_2/b_1. If W equals 1, b_2/b_1 is unity and n_2/n_1 follows the usual Boltzmann law; this result is a natural consequence of the assumption that radiative and collisional transition probabilities have values characteristic of thermodynamic equilibrium at the temperature T. Under interstellar conditions visible light corresponds to radiation at 10,000°K diluted by a factor, W, about equal to 10^{-15}. The terms in W can usually be neglected under interstellar conditions, and b_2/b_1 takes the simple form

$$\frac{b_2}{b_1} = \frac{1}{1 + A_{21}/n_e \gamma_{21}} \tag{4-39}$$

Exactly as in equation (4-34) the relative deviation from thermodynamic equilibrium is expressed in terms of two de-excitation probabilities, one of which is unaccompanied by any reverse process. Depending on the relative values of these two probabilities the population ratio will deviate far from or remain very close to its value in thermodynamic equilibrium at the kinetic temperature T. For example, if a permitted radiative transition is possible down from an excited level at a few eV, A_{kj} will be of order 10^8 sec^{-1}. Since γ_{kj} rarely exceeds 10^{-6} [see equation (4-22) and Table 4.2], it is evident $A_{kj}/n_e \gamma_{kj}$ will exceed $10^{14}/n_e$ and is enormously greater than unity for the values of n_e

to be expected in interstellar space. Evidently such levels will be almost completely depopulated under interstellar conditions.

For excited levels within the ground terms downwards transitions are forbidden for electric dipole radiation, and A_{kj} is enormously reduced. As a result, such levels are not much depopulated and may even approach their equilibrium population density relative to the ground state. An example is provided by the $^3P_1$ level of O III, for which the data in Table 4.2 give the result that at a kinetic temperature of 10,000°,

$$\frac{A_{kj}}{n_e \gamma_{kj}} = \frac{2400}{n_e} \qquad (4\text{-}40)$$

For normal interstellar densities the upper ^3P levels are strongly depopulated in H II regions. At electron densities of 10^3 cm^{-3} or more, these levels begin to approach their thermodynamic values, and upwards transitions from these levels to the $^1D_2$ level, for example, which were ignored in Table 4.2, would need to be considered.

The same analysis may be applied to excitation of the first excited rotational level of the CN molecule, discussed in Section 2.3. If the spontaneous radiation probability A_{kj} from this excited level is as large as 7×10^{-6} sec^{-1}, given by an approximate theoretical calculation, collisions can apparently not account for the observed excitation [27]. Black body radiation at a wavelength of 0.26 cm, corresponding to a temperature of 3°K, would explain the observations. Such a universal radiation field, which is also indicated by direct measurements at wavelengths up to 7 cm, is of very central interest for cosmogony [28] but apparently has little other effect on the interstellar medium.

For transitions at longer radio wavelengths A_{kj} becomes much less, tending to give b_2/b_1 nearly equal to 1. On the other hand, $h\nu/kT$ becomes very small compared to unity while the observed value of W may be comparable with unity. If we expand the exponents in equation (4-38), and assume that $n_H \gamma_{21}$ much exceeds A_{21}, we obtain for the correction factor appearing in the integrated atomic absorption coefficient, s, [equation (2-28)]

4. INTERACTIONS AMONG INTERSTELLAR PARTICLES

$$\left\{\frac{b_2}{b_1} + \frac{kT}{h\nu}\left(1 - \frac{b_2}{b_1}\right)\right\} = \frac{1 + \dfrac{A_{21}}{n_H \gamma_{21}}\dfrac{kT}{h\nu}}{1 + W\dfrac{A_{21}}{n_H \gamma_{21}}\dfrac{kT}{h\nu}} \qquad (4\text{-}41)$$

We have replaced n_e by n_H, since excitation of the hyperfine levels of H, for example, is primarily caused by collisions with other H atoms. If we insert numerical values for the 21-cm line, with T equal to $100°$, and γ_{21} evaluated from Table 4.3, we find that for small W the absorption coefficient is increased above its equilibrium value by the factor $1 + 1/(37 n_H)$. The correction is negligible for large n_H, but begins to be appreciable if n_H is as small as 0.1. For the stronger OH lines A_{21} is greater by a factor 10^4, and if γ_{21} is assumed to be about the same as for H, the correction factor for absorption should be $1/W$. As a result, on this idealized model the absorption coefficient would be inversely proportional to the mean radiation temperature, T_R, rather than to the kinetic temperature, T.

As noted in Section 2.2, observational evidence indicates that other mechanisms are probably responsible for the excitation of the OH levels. An additional mechanism which may be dominant for H at low density is scattering of $L\alpha$ photons, since the absorbing electrons may jump up from one hyperfine state and down to another. In a quiescent medium, this process leads indirectly to excitation of the hyperfine levels in accordance with the Boltzmann law at the kinetic temperature [29], but if large-scale motions are present the results are less clear.

4.3 Ionization and Dissociation

In a steady state n_{r+1}/n_r, the ratio of number densities for atoms in the stages of ionization $r + 1$ and r, is determined by the condition that the number of ionizing transitions per cm^3 per sec equals the corresponding number of recombinations. An equation similar to (4-31) may be written down for this case. The rate of energy absorbed per cm^3 in ionizing transitions from atoms of type X in energy level j in ionization stage r is $n_j(X_r)s_\nu cU_\nu d\nu$ integrated over all frequency; to obtain the number of such transitions the integrand must be divided by $h\nu$, the energy of each such transition. We define a rate coefficient, or radiative ionization probability β_{rjf}, for these transitions by the equation

$$\beta_{rjf} = c \int_{\nu_{rj}}^{\infty} s_\nu U_\nu \frac{d\nu}{h\nu} \qquad (4\text{-}42)$$

where ν_{rj} is the frequency corresponding to the energy required for ionization from level j. Similarly to β_{rjf} we define a rate coefficient ζ_{rjf} for ionization by energetic particles.

The rate coefficient α_{rj} for radiative capture of free electrons in level j of an atom in stage of ionization $r+1$ is defined in equation (4-27). Under interstellar conditions, ionization generally occurs from the ground level ($j = 1$), and the equation of steady state for ionization equilibrium becomes

$$n(X_r)(\beta_{r1f} + \zeta_{r1f}) = n_e n(X_{r+1}) \sum_j \alpha_{rj} = n_e n(X_{r+1})\alpha_r \quad (4\text{-}43)$$

where α_r is the total recombination coefficient for an atom in stage of ionization $r + 1$ [see equation (4-28)]. Dependence of α_{rj} on excitation level of the ions X_{r+1} is ignored in this expression. We shall generally omit the subscript r from these rate coefficients whenever the ion being referred to is clear; similarly the subscript 1 will be dropped from β_{1f} and ζ_{1f}, since all ionizing transitions considered here are from the ground level.

Ionization of hydrogen by energetic particles (sometimes called "suprathermal") may be of particular importance in those H I regions where the density is somewhat less than the average. According to equation (4-29) the cross section varies inversely as the square of the velocity, and hence the cosmic rays of lower energy make the larger contribution to ζ_f, the rate coefficient for ionization by energetic particles. Unfortunately the number of energetic particles in interstellar space at energies much below 10^9 eV is quite uncertain, as noted in Section 2.4. We can obtain upper and lower limits for ζ_f from the following arguments. If the

4. INTERACTIONS AMONG INTERSTELLAR PARTICLES

minimum possible corrections are made to the observed cosmic ray fluxes measured at the Earth, these fluxes show a peak at about 100 MeV. Integration of σ_{HR} over this minimum spectrum and multiplication by a factor 5/3 to allow for additional ionization produced by secondaries yields [30] the following minimum value for ζ_f, which we denote by ζ_E,

$$\zeta_E = 6.8 \times 10^{-18} \text{ sec}^{-1} \qquad (4\text{-}44)$$

Unless the cosmic rays observed at the Earth are mostly confined to the neighborhood of the solar system, it does not seem possible for ζ_f to be significantly less than this value.

To obtain an upper limit on ζ_f we assume that a third of the kinetic energy observed in type I supernova envelopes is available in the form of 2 MeV protons which permeate the Galaxy, a possibility discussed in more detail in Chapter 5. From the power density P_s in equation (5-27) and the energy loss of 36 eV per ion pair (see Section 4.1) we obtain for this upper limit, which we denote by ζ_s

$$\zeta_s = \frac{1.0 \times 10^{-15}}{\bar{n}_H} \text{ sec}^{-1} = 1.17 \times 10^{-15} \text{ sec}^{-1} \qquad (4\text{-}45)$$

where $\bar{n}_H$, the mean value of n_H in the solar neighborhood, has been set equal to 0.85 cm^{-3}, again increasing somewhat the galactic value of 0.5 cm^{-3} to allow for concentrations in the local spiral arms. While equation (4-45) does not give a firm upper limit for ζ_f, an ionization rate much larger than this value would seem to require some additional source of energy, since ζ_s begins to approach the limit permitted by the energy sources discussed in Section 5.1. If, for example, most of the energy from supernovae is assumed to be in a form not yet observed, greater values of ζ_f are possible.

A variety of additional ionization and recombination processes are not considered here, including collisional ionization by thermal electrons, together with the inverse process of three-body recombination, and also ionizing transitions which leave the ion in an excited state, together with the corresponding inverse process of electron capture by an excited ion. In principle all such processes can be included in equation (4-43) if their rate coefficients are known. In practice there are so many different processes that can lead to recombination, through various intermediate excited atomic and molecular levels, that it is sometimes difficult to be sure that the most important processes have all been considered.

As in the previous cases a simple relationship between β_{rjf} and α_{rj} is obtained from the condition of detailed balancing in thermodynamic equilibrium, which implies that the number of radiative ionizing transitions from level j per cm^3 per sec equals the corresponding number of radiative captures on the same level. The expressions for these two rates are similar to those in equation (4-43), with several obvious modifications. The values of α_{rj}, n_e and $n(X_{r+1})$ are equal to those in the equivalent thermodynamic system, but $n_j(X_r)$ and β_{rjf} must be replaced by $n_j(X_r)^*$ and β_{rjf}^*, with the latter given by equation (4-42) with $4\pi B_\nu(T)/c$ replacing U_ν. If equations (2-10) and (2-12) are used to eliminate the ratio of atom densities, we obtain.

$$\beta_{rjf}^*(T) = \frac{g_{r+1,1}}{g_{rj}} e^{-h\nu_j/kT} f_e \alpha_{rj}(T) \qquad (4\text{-}46)$$

where f_e is again the partition function for free electrons, given in equation (2-13). Since excited states of the atom in stage of ionization $r+1$ have not been considered in equation (4-43), f_{r+1} has been replaced by $g_{r+1,1} \exp(-E_{r+1,1}/kT)$, where $g_{r+1,1}$ is the weight of the ground level; $h\nu_j$ equals $E_{r+1,1} - E_{rj}$.

Equation (4-43) is applied below to the ionization equilibrium first of interstellar H and then of other elements. The corresponding equation for the formation and association of molecules is also discussed.

Hydrogen Ionization

We consider first the ionization of hydrogen produced by "ultraviolet photons," defined here as any photon with an energy $h\nu$ greater than E_1, the binding energy of the hydrogen atom in the

4. INTERACTIONS AMONG INTERSTELLAR PARTICLES

ground level. The rate coefficient, ζ_f, for collisional ionization will be ignored at first; ionization by energetic particles will be considered later. The electron and proton densities, n_e and n_p, will be assumed equal, and n_H will denote the sum $n(\text{H I}) + n_p$, where $n(\text{H I})$ is the particle density of neutral H atoms, and n_p is the proton density. If also x denotes the fractional ionization, n_p/n_H, equation (4-43) yields

$$\frac{1-x}{x^2} = \frac{n_H \alpha}{\beta_f} \tag{4-47}$$

where the subscripts r have been dropped from the rate coefficients.

The density of interstellar H is so great that the absorption of ionizing radiation has a profound effect on the solution of equation (4-47). We again define an integrated atomic cross section, s, by equation (2-21), with the integral now extending from the series limit to infinity; the optical depth in the continuum may then be found from equation (2-30). If s is found from equations (2-23) and (2-25), with an oscillator strength of 0.41 for the Lyman continuum, and if we write approximately

$$\phi(v) = \frac{s_v}{s} = \frac{2v_1^2}{v^3} \tag{4-48}$$

we obtain

$$d\tau_v = 6.6 \times 10^{-18}(1-x)n_H \left(\frac{v_1}{v}\right)^3 dL \tag{4-49}$$

where dL is used here to denote an interval of distance along the ray path. Thus if x is small, and n_H is 1 per cm^3, the mean free path for a photon of energy slightly greater than 13.6 eV, corresponding to hv_1, is only 0.05 pc, and increases to 5 pc only for hv as great as about 60 eV, where presumably very little energy is radiated by even the hottest normal stars. As a consequence, when an appreciable fraction of the hydrogen is neutral in a region several parsecs across, the absorption of any ultraviolet radiation striking the region will be so complete that essentially all of the hydrogen will be neutral. The only way that radiation from an early-type star can ionize hydrogen over a large region is to keep the neutral density so low that the optical depth across the region is not too large. It follows that except possibly in regions of very low density interstellar hydrogen will be divided into two types of regions, the H II regions in which x is nearly unity and the H I regions in which $1 - x$ is nearly unity [31]. The transition layer between the two regions will have an optical thickness of about unity in the Lyman continuum and will be relatively thin.

We compute here the radius r_S of the H II region surrounding an early-type star. Let $L_v dv$ be the luminosity of the star in the frequency interval dv. Then U_{sv}, the energy density of starlight at a distance r from the star is given by

$$U_{sv} = \frac{L_v e^{-\tau_v}}{4\pi r^2 c} \qquad (4\text{-}50)$$

The distance r is assumed to be much larger than the stellar radius. If we multiply equation (4-50) by $4\pi r^2 c$ and differentiate, we obtain

$$\frac{1}{r^2}\frac{d}{dr}(r^2 c U_{sv}) = -(1-x)n_H s_v c U_{sv} \qquad (4\text{-}51)$$

where we have used equations (2-2) and (2-20). It is assumed that only hydrogen atoms produce a significant absorption of the ultraviolet radiation; the effects produced by helium are considered later.

In addition to the stellar radiation there is the diffuse ultraviolet radiation field, of intensity I_{Dv}, emitted by those electrons which are captured directly in the ground level, emitting a photon in the Lyman continuum. If $\mathscr{F}_{Dv}$ is the net flux of this radiation, per cm² per sec per frequency interval, then equation (2-1) of radiative transfer, expressed in spherical coordinates, gives, on integration over all solid angles,

$$\frac{1}{r^2}\frac{d}{dr}(r^2 \mathscr{F}_{Dv}) = -(1-x)n_H s_v c U_{Dv} + 4\pi j_{Dv} \qquad (4\text{-}52)$$

4. INTERACTIONS AMONG INTERSTELLAR PARTICLES

where j_{Dv} is the emissivity for this diffuse radiation in the Lyman continuum. This relation simply equates the divergence of the diffuse flux to the source term minus the sink. If now we add equations (4-51) and (4-52), U_{sv} and U_{Dv} on the right-hand side combine to give the total energy density, U_v, at frequency v. If we substitute from equation (4-47) for $1 - x$, and multiply the resultant equation by dv/hv, integrating over all frequency, the integral of $cs_v U_v dv/hv$ in the numerator cancels β_f in the denominator [see equation (4-42)] and we obtain, on multiplying through by $4\pi r^2$

$$\frac{d}{dr}\left\{4\pi r^2 \int_{v_1}^{\infty}(cU_{sv} + \mathscr{F}_{Dv})\frac{dv}{hv}\right\} = -4\pi r^2\left\{x^2 n_H^2 \alpha - 4\pi \int_{v_1}^{\infty}\frac{j_{Dv}\,dv}{hv}\right\}$$

(4-53)

The second term in brackets on the right-hand side is simply the total number of electron captures in the ground state per cm³ per sec, equal to $x^2 n_H^2 \alpha_1$, and may be combined with the first term. If we define $S_u(r)$ as the total number of photons flowing through a shell of radius r per second with a frequency greater than v_1 (the frequency at the Lyman limit) the quantity in brackets on the left-hand side equals $S_u(r)$, and we obtain

$$\frac{dS_u(r)}{dr} = -4\pi r^2 x^2 n_H^2 \alpha^{(2)} \qquad (4-54)$$

where, for general k,

$$\alpha^{(k)} \equiv \sum_{j=k}^{\infty} \alpha_j \qquad (4-55)$$

Equation (4-54) gives the physically reasonable result that the net number of ultraviolet photons streaming outwards from the star is decreased by the number of recombinations to excited levels taking place within the region. Such captures in the second quantum level or higher produce radiation longwards of the Lyman limit, which yields no further ionization, and promptly escapes from the H II region. Electrons captured in the ground level, on the other hand, emit an ultraviolet photon, which will promptly be absorbed again before it has gone very far. Not until the electron is captured in a high quantum level will at least some of the resultant photons escape. Thus for computing the reduction of $S_u(r)$, captures in the ground level may be ignored [32].

Equation (4-54) may now be integrated over dr out to the radius r_S, defined as the value of r at which $S_u(r)$ has decreased to zero; this distance is sometimes called the "Strömgren radius." If n_H is assumed constant and x is set equal to unity throughout the region, one finds

$$\frac{4\pi}{3} r_S^3 n_H^2 \alpha^{(2)} = S_u(0) \tag{4-56}$$

At small r the diffuse flux, F_{Dv}, is negligible, and $S_u(0)$ is equal to the number of ultraviolet photons radiated by the central star.

The partial recombination coefficient $\alpha^{(2)}$ in equation (4-56) may be evaluated for a hydrogenic atom with a nuclear change of Ze by use of the electron capture cross sections given in equation (4-25). We introduce the functions $\phi_k(\beta)$, defined by the relationship

$$\alpha^{(k)} = 2A_r \left(\frac{2kT}{\pi m}\right)^{1/2} \beta \phi_k(\beta) \tag{4-57}$$

where A_r is given in equation (4-26), and β is defined by

$$\beta \equiv \frac{hv_1}{kT} = \frac{157{,}000 Z^2}{T} \tag{4-58}$$

If the Gaunt factor, g_{fn}, is assumed constant, the slowly varying functions $\phi_k(\beta)$ may be expressed, with use of equations (4-27) and (4-28), in terms of exponential integrals. Table 4.4 gives values of $\phi_1(\beta)$ and $\phi_2(\beta)$ computed with a series expansion for g_{fn} [33]. The errors in ϕ_1 and ϕ_2 resulting from the approximate evaluation of g_{fn} are believed to be less than 3% at temperatures less than 16,000°, with somewhat greater errors possible at

4. INTERACTIONS AMONG INTERSTELLAR PARTICLES

higher T. Inserting numerical values into equations (4-57) we obtain

$$\alpha^{(2)} = \frac{2.07 \times 10^{-11} Z^2}{T^{1/2}} \phi_2(\beta) \text{ sec}^{-1} \text{ cm}^{-3} \qquad (4\text{-}59)$$

The quantity $\alpha^{(1)}$ is usually designated more simply as α, the total recombination coefficient, [see equation (4-28)].

TABLE 4.4

Recombination Coefficient Functions ϕ_1, ϕ_2

$T(°K)$	31.3	62.5	125	250	500	1000
ϕ_1	4.68	4.36	4.04	3.71	3.38	3.05
ϕ_2	3.89	3.57	3.25	2.92	2.60	2.27
$T(°K)$	2000	4,000	8,000	16,000	32,000	64,000
ϕ_1	2.73	2.40	2.09	1.79	1.50	1.23
ϕ_2	1.96	1.64	1.34	1.06	0.80	0.59

Values of r_S computed from equation (4-56) for early-type stars are given in Table 4.5 together with the assumed values of the radius R [31] in solar units, and the stellar ultraviolet temperature

TABLE 4.5

Radii of H II Regions

Spectral type	T_c	$R/R_\odot$	$S_u(0)$	$r_s n_H^{2/3}$
O5	56,000°K	7.0	31×10^{48} sec^{-1}	100 pc cm^{-2}
6	44,000	6.7	9.0	66
7	36,000	6.3	2.7	44
8	30,000	6.0	0.77	29
9	25,000	5.6	0.19	18
B0	21,000	5.3	0.041	11
1	18,000	5.0	0.008	6.4

[34], T_c, defined as the temperature of a black body whose emitted flux shortward of the Lyman limit just equals that from the star. Also given in Table 4.5 are the total number of photons emitted by the star per second beyond the Lyman limit, computed from the Planck formula. The values of $r_S n_H^{2/3}$ have been computed for a kinetic temperature of $10^4\,°K$, for which the effective recombination coefficient given in equation (4-59) equals 2.6×10^{-13} cm^3 sec^{-1}. It may be noted that these values of r_S are independent of the specific way in which s_v varies with v, since for almost all the photons from even an O5 star the mean free path for absorption computed from equation (4-49) is small compared to r_S.

We assume that every recombining electron reaches the ground state before it absorbs an ionizing photon. Hence every recombination of an electron with a proton produces a photon of Lyman radiation in addition to other photons of longer wavelength. The Lyman photons are all strongly absorbed, and except for $L\alpha$ are soon degraded by fluorescence into $L\alpha$, a Balmer photon and other photons. Hence the total numbers of photons emitted initially in $L\alpha$ and in the Balmer radiation are each equal to $S_u(0)$. The Balmer photons all escape, since the optical depth is low, but the energy density of $L\alpha$ photons tends to build up appreciably because of the very large optical depth for scattering. If these $L\alpha$ photons escaped by pure coherent scattering through a shell containing about 3×10^{19} H atoms per cm^2 the energy density would be increased by about 10^6, the optical thickness for $L\alpha$, and would much exceed the density of other forms of interstellar energy. Four processes set a limit on the actual energy density of this radiation: (a) absorption by the solid particles [35] known to be present in H II regions (see Section 3.2); (b) replacement of a $L\alpha$ photon by two photons, either by initial radiative capture in the $2S$ level or by a collisional transition from the $2P$ to the $2S$ level, followed in either case by a downwards transition to the $1S$ level with the emission of two photons [36]; (c) escape of the photon because of Doppler shifts to frequencies of low s_v; such frequency shifts could be due either to thermal velocities [37] or systematic velocity differences across the gas;

4. INTERACTIONS AMONG INTERSTELLAR PARTICLES

(d) absorption by He atoms in the metastable 2^3S state [38]. The overall energy density of $L\alpha$ radiation to be expected and all the various effects which this radiation might produce have not yet been fully explored.

The effect of helium atoms and ions has been ignored in the above discussion of hydrogen ionization. Just as hydrogen degrades an ultraviolet photon into a photon of Lyman α and other photons, so helium converts each photon of ultraviolet radiation into a photon whose wavelength is either 584 or 304 Å, the resonance lines of He I and He II, respectively, in addition to producing other photons whose wavelengths are mostly greater than 912 Å, the wavelength at the Lyman limit. If these other photons are ignored, since most of them cannot ionize an H atom, the flux of photons capable of ionizing hydrogen is unaltered by the presence of helium, and the radius, r_S, of the ionized hydrogen region is still given by equation (4-56), with $S_u(0)$ unchanged. Since any radiation which can ionize helium can also ionize hydrogen, the radius of the region in which helium atoms will be ionized cannot much exceed r_S computed for hydrogen alone, even for the hotter stars. For stars of relatively late type the relative number of stellar photons capable of ionizing helium will be much less than the corresponding number for hydrogen, and the helium will be ionized only in a small region close to the central star, with a radius much less than r_S for hydrogen [39].

Outside of H II regions there are normally no ultraviolet photons from stars capable of ionizing atomic H. A small fraction of H atoms will be kept ionized by cosmic radiation. We may evaluate n_p/n_H from equation (4-43), neglecting β_f. If we assume that n_e equals $n_p + n_i$, where n_i is the particle density of the various heavy ions which are ionized by starlight (see below) we find

$$n_e = \tfrac{1}{2}n_i + \tfrac{1}{2}(n_i^2 + 4\zeta_f n_H/\alpha^{(2)})^{1/2} \qquad (4\text{-}60)$$

where $n(\text{H I})$ has been replaced by n_H in the square root, since n_p/n_H is assumed small. Again we have assumed that recombinations directly to the ground state may be ignored. For a temperature of 100° equation (4-59) and Table 4.4 give $\alpha^{(2)}$ equal to

6.9×10^{-12} cm^3 sec^{-1}; if we ignore n_i and determine ζ_f from equations (4-44) and (4-45) in turn, we find

$$n_e = \begin{cases} 1.0 \times 10^{-3} n_H^{1/2}, & \zeta_f = \zeta_E \\ 1.3 \times 10^{-2} n_H^{1/2}, & \zeta_f = \zeta_s \end{cases} \quad (4\text{-}61)$$

The fraction of H atoms ionized in this way is less than 0.1 if n_H exceeds 10^{-2} cm^{-3}. As we shall see below, this density of electrons from hydrogen may substantially exceed the density of heavy positive ions. Thus the ionization level produced by cosmic rays, although relatively small, may provide a dominant influence on the ionization equilibrium of H I regions.

Ionization of Sodium and Calcium

The ionization equilibrium of sodium and calcium atoms has been analyzed [40] in connection with the observations of the Na I and Ca II lines in absorption. The observational data on individual atomic densities, discussed in Section 2.3, are not sufficient to permit any definitive conclusions, but an attempt has been made to see if the results are generally consistent with theoretical expectations based on the same chemical composition observed in the stars and in planetary nebulae.

From a theoretical standpoint one difficult problem is to estimate β_f, the ionization probability from the ground state. In particular, the mean interstellar radiation density, U_ν, is quite uncertain in the far ultraviolet, though observations from space vehicles are beginning to fill this gap. Quantum-mechanical calculations of s_ν have been carried through for a number of astrophysically important atoms, and these have been combined with estimates of the radiation field to compute β_f for various atomic species [40]. The recombination coefficients, α_j, for the upper levels can be computed relatively accurately from the hydrogenic formula (4-25); those for the lower levels, which are generally less important, can be estimated from the corresponding β_f, with the use of equation (4-46). Resultant values of β_f/α (40), which in accordance with equation (4-43) is equal to $n_e n(X_{r+1})/n(X_r)$, are summarized in Table 4.6. For atoms whose ionization

4. INTERACTIONS AMONG INTERSTELLAR PARTICLES

TABLE 4.6

Theoretical Values of β_f/α

Atom	χ_r	Temperature (°K)		
		10^2	10^3	10^4
C I (H I)	11.3	14	63	320
Na I	5.1	0.68	3.3	24
Na II	47.3	0.018	0.079	0.39
Ca I	6.1	230	1200	8200
Ca II	11.9	0.071	0.31	1.5
Ca II (H I)	11.9	0.025	0.11	0.51

potential, χ_r, is not far below 13.6 eV some values, labeled H I, were computed with U_v equal to zero shortwards of the Lyman limit. As expected, the values tend to be greater for the lower excitation potentials. On the other hand the values for Na I and Ca II are relatively low, because most of the oscillator strength available from the ground level (limited by the f-sum rule) goes into the strong resonance doublets, and the f value for absorption in the continuum is about 0.01. The value of β_f is reduced accordingly. Computations of β_f based on a somewhat different radiation field give [41] values which are about the same for the lower χ_r, but which for larger χ_r are from two to four times higher than those used in Table 4.6. Even for Na II, for which the ionization probability is least, the value of β_f is about 10^{-13} sec^{-1}, some 2 orders of magnitude greater than ζ_s. Hence ionization of sodium and calcium atoms by cosmic rays can be ignored.

To compute the ratio of ionized to neutral atoms for each element we must know not only β_f/α but also the electron density, which depends on n_i, the density of heavy positive ions, as well as on ζ_f. In H I regions the ratio n_i/n_H may be assumed roughly constant and equal to the relative number of elements with a first ionization potential, χ_1, less than 13.6 v. For such elements the ratio of neutrals to ions will generally be small for conditions of interest, and except for Ca and similar elements the second ionization potential, χ_2, exceeds 13.6 V, and there will be very few atoms in the doubly ionized state. The value of n_i/n_H may then be determined from the relative abundances [42, 43] for the fourteen most abundant elements, given in Table 4.7, giving n_i/n_H equal to 4.9×10^{-4} in H I regions. In H II regions n_e equals n_p and is closely equal to n_H.

If $n(\text{Na II})$ and $n(\text{Ca III})$ are assumed to be about equal to n_{Na} and n_{Ca}, the total atom densities of Na and Ca including all ionization stages, then equation (4-43) and (4-60) may be combined with the composition in Table 4.7 to give $n(\text{Na I})$ and $n(\text{Ca II})$ as a function of n_H. These relationships become very simple in two cases. In H II regions n_e may be set equal to n_H, and equation (4-43) gives for Na and Ca, respectively,

$$n(\text{Na I}) = 8.3 \times 10^{-8} n_H^2 \text{ in H II} \qquad (4\text{-}62)$$

$$n(\text{Ca II}) = 1.1 \times 10^{-6} n_H^2 \text{ in H II} \qquad (4\text{-}63)$$

In H I regions if we set ζ_f equal to ζ_s, then n_i is negligible compared to n_e provided n_H does not exceed 100 per cm^3; equation (4-61) may then be used for n_e, and if we assume that T equals 100°K equation (4-43) now yields

$$n(\text{Na I}) = 3.8 \times 10^{-8} n_H^{3/2} \text{ in H I}, \zeta_f = \zeta_s \qquad (4\text{-}64)$$

$$n(\text{Ca II}) = 8.4 \times 10^{-7} n_H^{3/2} \text{ in H I}, \zeta_f = \zeta_s \qquad (4\text{-}65)$$

TABLE 4.7

Relative Cosmic Abundances of the Elements

Element	H	He	C	N	O	Ne	Na
Relative No. n, of atoms	10^6	1.0×10^5	400	110	890	500	2.0
First ionization potential χ_1(eV)	13.6	24.6	11.3	14.5	13.6	21.6	5.1

Element	Mg	Al	Si	S	Ar	Ca	Fe
Relative No. n, of atoms	25	1.7	32	22	7.8	1.6	3.7
First ionization potential χ_1(eV)	7.6	6.0	8.1	10.4	15.8	6.1	7.9

4. INTERACTIONS AMONG INTERSTELLAR PARTICLES

If ζ_f is set equal to ζ_E, n_i cannot be ignored, and equation (4-60) must be solved for n_e, in general, with n_i/n_H equal to the value 4.9×10^{-4} found above.

These results may be compared with the observed values discussed in Section 2.3. First we consider the low density found between the Sun and α Vir, where the mean n(Ca II) is 1.3×10^{-10} cm^{-3}. If an H II region is assumed, the r.m.s. n_H in this region equals about 0.01 cm^3. On the assumption of an H I region, n_H equals 0.003 and 0.016 for the higher and lower values of ζ_f, respectively. Apparently the hydrogen density in this region is roughly about 2 orders of magnitude below its average value in the solar neighborhood.

At the other extreme is the large value of 3×10^{-8} cm^{-3} found for the mean N(Na I) in the line of sight to χ^2 Ori. The corresponding r.m.s. n_H is 0.6 for H II and 0.9 and 3, respectively, for H I with maximum and minimum ζ_f. If the high-density cloud extends over only a fraction F_c of the 1200 pc line of sight to this star, n(Na I) is increased by $1/F_c$ and n_H will be increased by a smaller factor. If the cloud is assumed to be 25 pc across, F_c equals 0.02 and the cloud density, n_H, is about 4 cm^{-3} on the H II hypothesis, and between 12 and 30 cm^{-3} for H I. The lower n_H in H II regions results from the substantially higher electron density than in H I regions, which tends to increase n(Na I)/n(Na II) and n(Ca II)/n(Ca III), a tendency only partly offset by the high T and consequently lower α in H II regions.

Next we apply equations (4-62) and (4-63) to the mean observed values of n(Na I) and n(Ca II) given in equation (2-61), which for Na represents unreddened stars. For Na we find that the r.m.s. n_H equals 0.16 cm^{-3} on the assumption of line formation in an H II region, with somewhat greater values for H I. Quite possibly both types of regions contribute significantly to absorption line formation. From the Ca lines, on the other hand, the r.m.s. values of n_H are found to be less by about a factor 1/4; it is evident from a comparison either of equations (4-62) and (4-63) or of (4-64) and (4-65) that n(Ca II) should exceed n(Na I) by a factor of about 30, while in fact these two particles densities are found to be equal [see equation (2-61)].

This apparent underabundance of Ca relative to Na apparently cannot be fully explained by errors in the assumed rate coefficients, although this possibility cannot be completely excluded. The decrease of n(Na I/N(Ca II) with increasing cloud velocity, shown in Table 2.2, suggests that the relative abundances are normal in the high-speed clouds. From the observed ratios in Table 2.2 and the values of β_f/α for H I regions in Table 4.6, we find that n_{Na}/n_{Ca} falls from 100 in the low-velocity clouds to 10 in the high-velocity clouds. One suggested mechanism for explaining this change with cloud velocity is that the calcium atoms are preferentially locked up in the grains, which in the high-speed clouds have partially evaporated somehow during the acceleration process.

Finally we discuss the value of n(Na I) in the reddened clouds, for which equation (2-63) is applicable. If we combine this equation with equation (4-64), for an H I region with ζ_f equal to ζ_s, and assume that the average value of n_H within the clouds equals $\bar{n}_H F_c^{-1}$, cm^{-3}, we find that if $\bar{n}_H$ exceeds 0.26 cm^{-3} no solution is possible with F_c less than unity; i.e., the electron density corresponding to the average value of n_H is high enough in this case so that the density of Na I atoms in equilibrium exceeds the mean number observed. If the ionization rate produced by cosmic rays is really so high, some explanation for this discrepancy must be found. Large errors in β_f/α seem somewhat unlikely. Another possibility is that the heavy atoms are mostly locked up in the grains, and that the free sodium atoms in the gas represent only a small percentage of the sodium atoms present in the interstellar medium, with the calcium atoms representing an even smaller percentage.

If ζ_f is set equal to ζ_E, the minimum ionization rate, the same assumptions yield

$$F_c = \begin{cases} 0.42, \bar{n}_H = 0.85 \\ 0.12, \bar{n}_H = 0.50 \end{cases} \quad (4\text{-}66)$$

giving a value for n_H within the clouds between 2.0 and 4.2 cm^{-3}. This corresponds to rather a slight concentration of material

4. INTERACTIONS AMONG INTERSTELLAR PARTICLES

within the clouds, and a smaller value of F_c has been used in Table 3.4. If the reddened clouds are assumed to be all H II regions, equations (2-63) and (4-62) give a r.m.s. proton density of about 0.3 cm^{-3}, a not unreasonable value. However, it seems likely that only a small fraction of the obscuring clouds are H II regions, and strong absorption in the Na D lines by a few such clouds can pobably not explain the relatively good correlation of $n(\text{Na I})$ with color excess that appears to be present.

Evidently the theoretical discussion of ionization equilibrium in the interstellar medium is quite inconclusive at the present time because of uncertainty in n_e. New types of observations seem to be required.

Dissociation Equilibrium for Molecules

The relative numbers of atoms and molecules in a steady state are given by an equation analogous to (4-43). While not much quantitative information is available on rate coefficients concerned with molecules, a few conclusions have been tentatively established.

The H_2 molecule cannot be formed by radiative association of two interstellar H atoms, since vibrational–rotational transitions are forbidden in this homonuclear molecule; by symmetry the center of charge of the two nuclei always coincides with the center of mass, the electric dipole moment vanishes, and radiation in the vibrational–rotational lines is produced by the much weaker electric quadrupole or magnetic dipole moments. Formation of the H_2 molecule can probably occur on the surfaces of grains, a process discussed in Section 4.6. Destruction of the H_2 molecule can occur in several ways. In H II regions photons more energetic than 15.42 or 14.67 eV can produce ionization or dissociation directly [19], and the lifetime of the H_2 molecule in such a region is very short [32]. In H I regions destruction occurs by more indirect routes. For example a molecule absorbing a photon in one of the absorption lines shortward of 1108 Å will jump down again to the ground electronic level, emitting another photon. However, the vibrational quantum number can change appreciably in each transition, and if this quantum number in the ground electronic level is 14 or more the molecule dissociates [19]; the transition probabilities for such transitions indicate that about 0.1 of the line absorptions will produce dissociation in this way [44]. Under typical interstellar conditions an H_2 molecule will dissociate in this way in about 600 years, on the average, provided that line absorption does not weaken the intensity in the molecular lines. The equilibrium abundance of H_2, depending on the formation rate, is discussed in Section 4.6.

The equilibrium of CH and CH^+ molecules has been analyzed quantitatively [45], with rate coefficients estimated for the following processes: photodissociation, radiative attachment, photoionization, radiative recombination, and dissociative recombination (in which an electron combines with CH^+, with free C and H atoms resulting). Despite some uncertainties it appears that the mean densities given in equation (2-64) are greater than can be explained by these processes if n_H is less than 10^3 cm^{-3}. Again, the assumption of molecule formation on the surface of the grains, discussed in Section 4.6, seems a plausible resolution of the discrepancy. In the case of OH it appears [46] that the charge transfer reaction

$$O + H_2 + \Delta \to OH + H \tag{4-67}$$

where Δ represents an energy of about 0.10 eV, can produce appreciable OH if H_2 is abundant; for this process to be important the cloud must be heated to a temperature of at least 1000°K, which may be anticipated when two clouds collide (see Section 5.1).

4.4 Kinetic Temperature

The kinetic temperature of the gas in a steady state is determined by the condition that the total kinetic energy gained per cm^3 per sec, which we denote by Γ, is equal to the corresponding energy lost per cm^3 per sec, denoted by Λ. In general both Γ and Λ will depend on the temperature, T, as well as on the particle

4. INTERACTIONS AMONG INTERSTELLAR PARTICLES

density; the value of T at which these two functions are equal will be an equilibrium temperature, T_E. We postpone until later the consideration of possible differences in T between different types of particles, such as electrons and H atoms, and also the consideration of whether the equilibrium is thermally stable.

We denote by subscripts ζ, η the interacting particles responsible for each contribution to the total Γ and Λ. The condition that $\rho(\Gamma - \Lambda)$, the energy input per gram per second, equals the corresponding rate of increase of thermal energy, plus the work done by the gas, gives the result [47]

$$\frac{d}{dt}\left(\frac{3}{2}nkT\right) - \frac{5}{2}kT\frac{dn}{dt} = \sum_{\zeta,\eta}(\Gamma_{\zeta\eta} - \Lambda_{\zeta\eta}) = \Gamma - \Lambda \quad (4\text{-}68)$$

where n is the total number of free particles per cm^3 in the interstellar gas. We assume that all components of the gas have the same kinetic temperature, T, and ignore heat conduction. The left-hand side of equation (4-68) is $\rho T dS/dt$, where S is the entropy per gram of the interstellar medium, assumed to be a perfect monatomic gas. We may define a "cooling time," t_T, by the relation

$$\frac{d}{dt}\left(\frac{3}{2}nkT\right) = -\frac{3nk(T - T_E)}{2t_T} \quad (4\text{-}69)$$

In the simple case where t_T and T_E are constant, $T - T_E$ approaches zero as $\exp(-t/t_T)$. When the density is constant, t_T equals the ratio of the excess of energy density over its value in equilibrium to the net cooling rate, $\Lambda - \Gamma$.

In order to compute T_E for interstellar conditions we turn now to a derivation of formulae for Γ and Λ for some of the dominant processes. A primary mechanism for heating the interstellar gas is photoelectric ionization of neutral atoms. Let E_2 denote the kinetic energy of the ejected electron. Not all this energy can be counted as a gain, however, since in a steady state each photoionization must be offset by a corresponding capture of a free electron, whose kinetic energy we denote E_1. Let n_i denote the particle density of the ionized atoms. Since the number of captures to level j of the neutral atom per cm^3 per sec is given by $n_e n_i \langle w\sigma_{cj} \rangle$, the final net gain associated with electron-ion recombinations, which we denote by Γ_{ei}, is given by

$$\Gamma_{ei} = n_e n_i \sum_j (\langle w\sigma_{cj} \rangle \bar{E}_2 - \langle w\sigma_{cj} E_1 \rangle) \qquad (4\text{-}70)$$

where $\langle \ \rangle$ again denotes an average over a Maxwellian distribution, while $\bar{E}_2$ denotes that an average is taken over all the ionizing photons. Under interstellar conditions essentially all photoionizations take place from the ground level, and hence $\bar{E}_2$ is independent of j; with use of equations (4-27) and (4-28) equation (4-70) may then be written

$$\Gamma_{ei} = n_e n_i \{\alpha \bar{E}_2 - \tfrac{1}{2} m \sum_j \langle w^3 \sigma_{cj} \rangle\} \qquad (4\text{-}71)$$

It will be noted that these equations for Γ_{ei} do not depend on β_f or on the radiation density. As may be seen from equation (4-43), when n_p is about equal to n_H, and the fraction of neutrals is accordingly low, an increase of radiation density and hence of β_f decreases $n(\text{H I})$, but the number of ionizing transitions per cm^3 per sec is essentially unchanged, being fixed by the recapture rate $n_e n_p \alpha$.

For energy loss in interstellar space the primary mechanism is inelastic collisions between particles. In collisions between electrons and ions the number of excitations from level j to level k per cm^3 per sec will be $n_e n_{ij} \gamma_{jk}$, where n_{ij} is the number of ions in level j and γ_{jk} is the appropriate rate coefficient [see equation (4-4)]. The kinetic energy lost by the colliding electrons will be $E_k - E_j$, or E_{jk}. The deexciting collisions will produce an offsetting energy gain, and if we sum over all transitions between all levels the net rate of energy loss in electron–ion collisions, which we denote by Λ_{ei}, will be given by

$$\Lambda_{ei} = n_e \sum_{j > k} E_{jk}(n_{ij} \gamma_{jk} - n_{ik} \gamma_{kj}) \qquad (4\text{-}72)$$

Under most interstellar conditions which we shall consider all the ions in question will be in the ground level, and the sum over j can

4. INTERACTIONS AMONG INTERSTELLAR PARTICLES

be omitted, with n_{ij} for the ground level set equal to n_i. Similar equations hold for Λ_{ea}, Λ_{Ha}, and Λ_{Hm}, the losses resulting from excitation of neutral atoms by electrons, of ions by H atoms, and of molecules by H atoms, respectively. In addition to such processes kinetic energy will also be lost in free–free transitions, for which Λ_{ff} equals the rate of emission, ε_{ff}, given in equation (2-47); the energy gained by free–free absorption of starlight is completely negligible.

H II Region of Pure Hydrogen

We apply these concepts first to an H II region in which no atoms other than hydrogen are assumed present. To simplify the situation we ignore electron impact; in actual H II regions the excitation energy of 10.2 eV required for the level $n = 2$ is so much greater than the mean electron energy of about one eV that the number of such excitations is negligible. At the higher temperature found below for pure hydrogen nebulae this neglect is not always valid, but provides a useful introduction to the subject. On this basis the energy gain, Γ_{ep}, is offset only by the free–free emission rate, ε_{ff}. The temperature T is therefore determined by equating Γ_{ep}, obtained from equation (4-71), to ε_{ff} from equation (2-47).

The computation of $\bar{E}_2$ in equation (4-71) is complicated by two effects within an H II region, the absorption of the stellar radiation [see equation (4-50)] which varies with frequency, and the presence of the diffuse ultraviolet radiation field. To facilitate the analysis we consider here two relatively simple situations. First we consider Γ_{ep} and the resultant value of the equilibrium temperature T_E, near the exciting star, where both the diffuse radiation and the absorption of the stellar radiation can be ignored. Next we consider the average value of Γ_{ep} for the H II region as a whole; the diffuse radiation does not affect this average, and the absorption can be taken into account, since all photons in the Lyman continuum are absorbed somewhere in the nebula.

As r, the distance from the exciting star, decreases, the stellar radiation field increases as $1/r^2$ and becomes large compared to the diffuse ultraviolet radiation. Hence for r/r_S small we may neglect the diffuse radiation field and compute $\bar{E}_2$ by averaging $hv - hv_1$ over U_v/hv, the number density of photons in the stellar radiation field only; as before v_1 denotes the frequency at the Lyman limit. If we assume that U_v is dilute blackbody radiation at the color temperature, T_c, we obtain

$$\bar{E}_2 = \frac{\int_{v_1}^{\infty} h(v - v_1) s_v B_v \, dv/v}{\int_{v_1}^{\infty} s_v B_v \, dv/v} \equiv kT_c \psi_0(\beta_c) \qquad (4\text{-}73)$$

If we substitute equation (2-4) for B_v into this equation, we find that the function $\psi_0(\beta_c)$ is one of the functions $\psi_m(x)$ defined by

$$\psi_m(x) + x = \int_x^{\infty} \frac{y^m \, dy}{e^y - 1} \left[\int_x^{\infty} \frac{y^{m-1} \, dy}{e^y - 1} \right]^{-1} \qquad (4\text{-}74)$$

If we expand $(1 - e^{-y})^{-1}$ in each denominator in a series of ascending powers of e^{-y}, $\psi_m(x)$ can be evaluated analytically; numerical values of $\psi_0(\beta_c)$ are given in Table 4.8.

TABLE 4.8

Functions $\psi_m(\beta_c)$ for Mean Photoelectron Energy

Temperature, T_c (°K)	4,000	8,000	16,000	32,000	64,000
β_c	39.50	19.75	9.88	4.94	2.47
$\psi_0(\beta_c)$	0.977	0.959	0.922	0.864	0.775
$\psi_3(\beta_c)$	1.051	1.101	1.199	1.380	1.655

For the average value of $w^3 \sigma_{cj}$, the second quantity in equation (4-71) that must be evaluated, we write

$$\sum_{j=k}^{\infty} \langle w^3 \sigma_{cj} \rangle = \frac{2A_r}{\pi^{1/2}} \left(\frac{2kT}{m} \right)^{3/2} \beta \chi_k(\beta) \qquad (4\text{-}75)$$

where A_r and β are defined in equations (4-26) and (4-58). Table 4.9 gives values of $\chi_1(\beta)$ and $\chi_2(\beta)$ computed [33] with a series expansion for g_{fn}. The accuracy is about the same as that for ϕ_1 and ϕ_2 in Table 4.4.

4. INTERACTIONS AMONG INTERSTELLAR PARTICLES

If now equations (4-73) and (4-75) are substituted into equation (4-71) we obtain, inserting numerical values,

$$\Gamma_{ep} = \frac{2.85 \times 10^{-27} n_e n_p}{T^{1/2}} \{T_c \psi_0(\beta_c)\phi_1(\beta) - T\chi_1(\beta)\} \quad (4\text{-}76)$$

where Γ_{ep} denotes the value of Γ resulting from interactions between electrons and protons. Equating Γ_{ep} to the free-free

TABLE 4.9

Energy Gain Functions χ_1, χ_2

$T(°K)$	31.3	62.5	125	250	500	1000
χ_1	4.24	3.90	3.56	3.23	2.90	2.58
χ_2	3.46	3.12	2.78	2.45	2.12	1.80
$T(°K)$	2,000	4,000	8,000	16,000	32,000	64,000
χ_1	2.26	1.95	1.65	1.37	1.10	0.84
χ_2	1.49	1.20	0.92	0.67	0.46	0.30

radiation rate in equation (2-47), with $\bar{g}_{ff}$ set equal to unity, yields for T_E

$$T_E = \frac{\phi_1(\beta)\psi_0(\beta_c)}{\chi_1(\beta) + 0.5} T_c \quad (4\text{-}77)$$

For T equal to 16,000°, $\phi_1/(\chi_1 + 0.5)$ equals 0.96, according to the values in Tables 4.4 and 4.9, and decreases somewhat as T increases. For T_c between 32,000° and 64,000° the value of $\psi_0(\beta_c)$ varies between 0.8 and 0.9, and T_E for an H II region of pure hydrogen, near the exciting star, will vary between about 25,000° and 50,000°.

Next we consider the mean value of Γ_{ep} averaged over the entire ionized region. In this case detailed balancing must hold for the diffuse ultraviolet radiation. For each electron captured in the ground level, with emission of a photon of frequency v, there will be a photon of frequency v absorbed somewhere in the H II region. Hence the kinetic energy gained from absorption of the diffuse radiation is exactly equal to the loss; we conclude that captures to the ground state contribute nothing to the mean Γ_{ep} and may be ignored, together with the diffuse radiation which they produce. As a result, in equation (4-71) both terms on the right-hand side may be computed excluding captures to the ground level, exactly as in the computation of r_S in the previous section. Hence we must take k equal to 2 in equation (4-75).

In addition, when an average over the entire H II region is being considered E_2 is no longer given by equation (4-73). Since all ultraviolet photons (v greater than v_1) are absorbed somewhere in the region, the correct value of $\bar{E}_2$ is the total excess energy available from the star, divided by $S_u(0)$, the total number of ultraviolet photons radiated from the star. It follows that s_v must now be omitted from equation (4-73) and $\bar{E}_2$ becomes

$$\bar{E}_2 = kT_c \psi_3(\beta_c) \qquad (4\text{-}78)$$

Values of $\psi_3(\beta_c)$ appear in Table 4.8.

Combining these results gives for the mean Γ_{ep} an equation identical to (4-76), but with ϕ_2, χ_2, and ψ_3 replacing ϕ_1, χ_1, and ψ_0, respectively. For approximate results we equate the mean gain Γ_{ep}, to the free–free emission at the mean equilibrium temperature, which we denote by $\bar{T}_E$, and obtain,

$$\bar{T}_E = \frac{\phi_2(\beta)\psi_3(\beta_c)}{\chi_2(\beta) + 0.5} T_c \qquad (4\text{-}79)$$

The value of $\phi_2/(\chi_2 + 0.5)$ at 16,000° is 0.91, about 5% less than the corresponding value of $\phi_1/(\chi_1 + 0.5)$. For T_c between 32,000 and 64,000 ψ_3/ψ_0 varies between 1.6 and 2.1. Hence $\bar{T}_E$ is between 1.5 and 2 times the value of T_E near the exciting star; this temperature increase results from the selective extinction of the stellar radiation at the lower ultraviolet frequencies, producing a marked increase of $\bar{E}_2$ with increasing distance from the star. In an actual situation dynamical processes, which are ignored here, will change the uniform density assumed in these calculations, and impurities will reduce T_E, as discussed below.

H II Region with Impurities

In an actual H II region collisional excitation of the ions present provides a large loss of kinetic energy. Since the cross section for collisional excitation can exceed the radiative capture cross section by a factor of 10^5, such collisions provide a powerful cooling mechanism. Were it not for the relatively very low abundance of all atoms heavier than helium the interstellar gas would cool to a very low temperature even in H II regions.

For exciting an atom up to a different spectroscopic term the energy required is normally a few electron volts. At temperatures below 10,000° the mean electron energy is less than 1 eV, and the rate of such reactions falls off rapidly as the temperature decreases. These transitions provide a thermostatic mechanism that tends to keep the electron temperature below about 10,000°. The different fine-structure levels of a spectroscopic term are normally separated by a small fraction of an electron volt, and collisional excitation of these levels in the ground term is less sensitive to temperature for T greater than about 1000°. The energy radiated in the infrared by spontaneous downwards transitions from these fine-structure levels can therefore cool the gas to a relatively low temperature. In most H II regions the abundant atoms are singly ionized. O II has no fine structure levels in the ground term, but C II, N II, and Ne II all do (see Table 4.2). The abundance of N given in Table 4.7 is relatively low, but C and Ne contribute about equally to the cooling of the gas at low T. Around the hottest stars doubly ionized elements appear, and infrared radiation by O III becomes dominant.

Values of Λ_{ei} are shown in Figure 4.1, computed [12] for a gas with the impurities C II, N II, O II and Ne II present with the abundances in Table 4.7. The relevant energy levels and their collision strengths are listed in Table 4.2; the values of Ω used in Figure 4.1 differ slightly from those in Table 4.2. The sharp rise in Λ_{ei} shown in Figure 4.1, starting at a temperature of about 6000°, results mainly from collisional excitation of O II; the lowest excited terms of C II and Ne II have excitation energies exceeding 5 eV and contribute little at this temperature. The fine-structure levels account for the nearly flat section of the Λ_{ei} curve at low values of T. Also shown in the figure is Γ_{ep}, obtained from equation (4-76) for T_c equal to 32,000°, and ε_{ff} from equation

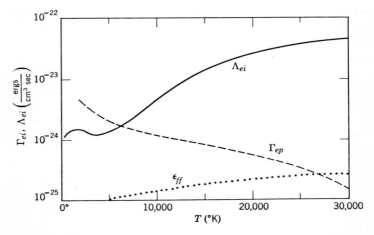

Fig. 4.1. Energy gain and loss rates in H II regions. The solid and dashed curves show Λ_{ei} [12] and Γ_{ep}, respectively, for electron and proton densities of 1 cm^{-3}. The stellar ultraviolet temperature, T_c, is taken to be 32,000°; C, N, O, and Ne are assumed singly ionized, with the relative abundances in Table 4.7. The dotted curve shows ε_{ff}. Redrawn from ref. 12.

(2-47); evidently free–free emission is quite negligible compared to radiation by the impurities. The equilibrium temperature obtained from Figure 4.1 is about 7000°K. Evidently a fivefold decrease in the impurity level would not increase the temperature much above 10,000°, but a fivefold increase would decrease T_E to a very low value. To a first approximation changes in the electron density do not affect T_E, since both Λ_{ei} and Γ_{ep} tend to vary as n_e^2. For electron densities as great as 10^3 cm^{-3}, however, the relative populations in the fine structure levels of the ground terms tend to approach their values in thermodynamic equilibrium [see equation (4-34)]. Hence at these higher densities the

4. INTERACTIONS AMONG INTERSTELLAR PARTICLES

contribution of the infrared emission to Λ_{ei} increases only linearly with n_e, and the equilibrium temperature rises to about 9000°K for the composition and T_c assumed in Figure 4.1.

Since equation (4-76) for Γ_{ep} is valid only near the center of the H II region, the temperature found from Figure 4.1 is strictly applicable only for low r/r_S. However, if the mean Γ_{ep} is taken for the region as a whole, $\psi_3 \phi_2$ is not very different from $\psi_0 \phi_1$, and the first term in equation (4-76) is not much changed; the second term is less important, since T is substantially smaller than T_c. Hence the mean Γ_{ep} averaged over the entire region will have nearly the same value as it does close to the center. Physically, the increased energy of the photoelectrons emitted for the H II region as a whole, resulting from the more energetic photons which are less strongly absorbed close to the star, is largely offset by the decrease in the recombination coefficient for the region as a whole, resulting from reabsorption of the diffuse flux.

The complex structure of an H II region can be investigated precisely by numerical computations, which can in principle take all effects into account. Extensive calculations including many of the relevant processes show [48] the effects produced by the absorption of the stellar radiation, with $\bar{E}_2$ increasing rapidly with increasing r. Ionization equilibrium, thermal equilibrium, and deviations of the stellar radiation from a blackbody curve were all included, as was the presence of helium, but all dynamical effects were ignored, as was the diffuse ultraviolet radiation. For a star of effective temperature, T_e, equal to 30,000°, the computed values of T_E ranged from 5000° near the center to 6000° at r equal to r_S; for a star with T_c equal to 50,000°, T_E ranged from 6000° up to about 10,000°.

The rate at which the gas temperature changes, when T is not equal to T_E, may be determined from the general equation (4-68). If the equilibrium temperature is 7000°K, as in Figure 4.1, then for values of T in the neighborhood of T_E the value of the radiative cooling time, t_T, defined in equation (4-69), is given roughly by

$$t_T = 2.0 \times 10^4/n_p \text{ years} \qquad (4\text{-}80)$$

H I Region

When the hydrogen is neutral, the gain resulting from electron captures and subsequent reionization is enormously reduced. The value of Γ_{ei}/n_e is proportional to the ion density, which is less by a factor 1/2000 in H I regions. In contrast the corresponding loss rate function per free electron, equal to Λ_{ei}/n_e, is not much changed. Since the O, Ne, and N atoms are neutral, the loss rate at high temperatures shown in Figure 4.1 is reduced to a lower value, determined by the lower excitation cross section for neutral atoms. However, the abundance of C^+ relative to hydrogen is the same in H I and H II regions, and at temperatures below 1000° the value of Λ_{ei}/n_e will be about the same in the two regions. Evidently the equilibrium temperature in H I regions tends to be relatively low.

Because of the low value of $n_e/n(\text{H I})$, processes involving neutral H also must be taken into account in computing T_E in H I regions. The value of Γ can be appreciably increased by cosmic-ray ionization of H atoms, a process contributing an amount Γ_{HR} to the kinetic energy per cm³ per sec. If $\langle E_h \rangle$ denotes the mean energy available for heating the gas, per free electron produced, and if ζ_f again denotes the probability per sec that a H atom will be ionized by energetic particles, then we have

$$\Gamma_{HR} = n(\text{H I})\langle E_h \rangle \zeta_f \tag{4-81}$$

We have seen in Section 4.1 that the mean energy lost in H gas by the energetic particles, per free electron produced, is about 36 eV. Of this amount 13.6 eV is required for the ionization energy of the electrons, and much of the rest is lost by collisional excitation of H atoms. Elastic collisions between the secondaries and the thermal electrons can transfer kinetic energy to the gas, elastic collisions with H atoms being relatively less important. However, equation (4-14) together with the cross sections for excitation of H indicate [30] that if n_e/n_H is less than 10^{-2}, electron excitation and ionization of H atoms will be the dominant energy loss rate until the electron energy falls below 10.2 eV, the threshold energy

4. INTERACTIONS AMONG INTERSTELLAR PARTICLES

for excitation. The last inelastic collision experienced by a secondary with more kinetic energy than 10.2 eV usually leaves it with an energy much less than this threshold value, and a detailed study [30] indicates that $\langle E_h \rangle$ equals 3.4 eV, giving a heating efficiency of about 9%. If n_e/n_H is greater than 0.01, however, the secondary electrons may lose a significant fraction of their energy directly to the thermal electrons, and the heating efficiency may increase somewhat.

With $\langle E_h \rangle$ set equal to 3.4 eV, and with ζ_f found from equations (4-44) and (4-45), respectively, equation (4-81) gives

$$\frac{\Gamma_{HR}}{n(\text{H I})} = \begin{cases} 3.7 \times 10^{-29} \text{ ergs/sec}, & \zeta_f = \zeta_E \\ 6.4 \times 10^{-27} \text{ ergs/sec}, & \zeta_f = \zeta_s \end{cases} \quad (4\text{-}82)$$

The value of Γ_{ei} corresponding to ionization of C may be obtained directly from equation (4-76) with $n(\text{C II})$ replacing n_p. The functions ϕ_2 and χ_2 are appropriate here because the inner shell with n equal to 1 is filled, and recombinations occur to the levels n equal to 2 or more. In the computations described below, the color temperature, T_c, in equation (4-76) has been set equal to 16,000°.

The chief loss processes that must be taken into account in H I regions are excitation of ions (primarily C^+) by electrons and excitation of ions and H_2 molecules by neutral H atoms. Rate coefficients for these processes may be obtained from Section 4.1. In particular, equations (4-19) and (4-22), together with the data in Table 4.2, give the information required for Λ_{ei}; similarly, equations (4-23) and (4-24), together with Table 4.3, give the rate coefficients needed for Λ_{Hi} and Λ_{Hm}. To take superelastic collisions into account, Λ_{Hm} must be computed from an equation corresponding to (4-72) for atom-molecule collisions, with the fraction of molecules in excited states computed from the relations in Section 4.2. Figure 4.2 shows theoretical values for Γ_{ei}, Γ_{HR}, Λ_{Hm}, and $\Lambda_{ei} + \Lambda_{Hi} + \Lambda_{Ha}$ computed in this way, with $n(\text{H I})$, n_e, and $n(\text{H}_2)$ set equal to 1, 10^{-2}, and 10^{-1} cm^{-3}, respectively. For this ratio of n_e to $n(\text{H I})$ the value of Λ_{ei} is about twice Λ_{Hi}.

In an actual situation the ionization level will vary with T. The loss rate Λ_{Ha} is relatively small for T less than 400°, but becomes important at higher T.

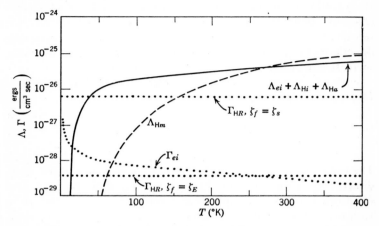

Fig. 4.2. Energy gain and loss rates in H I regions. The solid curve shows $\Lambda_{ei} + \Lambda_{Ha} + \Lambda_{Hi}$ for $n(H\ I)$ and n_e equal to 1 and 10^{-2} cm^{-3}, respectively; excitation only of C^+ and Si^+ ions and of O atoms is considered. The dashed curve shows Λ_{Hm} for $n(H_2)/n(H\ I)$ equal to 0.1. The dotted curves show Γ_{ei} and the upper and lower limits on Γ_{HR}.

TABLE 4.10

Temperature and Electron Density in H I Regions

$n(H\ I)$ (cm^{-3})	H ionization rate $= \zeta_E$		H ionization rate $= \zeta_s$	
	$T_E(°K)$	n_e (cm^{-3})	$T_E(°K)$	n_e (cm^{-3})
0.01	38	7.5×10^{-5}	860	2.7×10^{-3}
0.03	30	1.3×10^{-4}	250	3.0×10^{-3}
0.1	24	2.2×10^{-4}	120	4.2×10^{-3}
0.3	20	4.2×10^{-4}	70	6.3×10^{-3}
1	17	8.0×10^{-4}	45	1.0×10^{-2}
3	14	1.8×10^{-3}	33	1.6×10^{-2}
10	12	5.0×10^{-3}	27	3.0×10^{-2}

Values of the equilibrium temperature as a function of $n(\text{H I})$ for the higher and lower values of ζ_f are given in Table 4.10. Increase of n_e by cosmic ray ionization has been taken into account in accordance with equation (4-60), but cooling by molecular hydrogen has been ignored. Additional cooling processes which are usually negligible and which have been ignored are cooling by other molecular collision [49], involving either H_2 or such molecules as CH, NH, or OH, and inelastic collisions between H atoms and grains [50]. For values of n_e/n_H exceeding 0.01 the heating rate Γ_{HR} should exceed the value adopted, with a resulting increase of T_E, since $\langle E_h \rangle$ will exceed 3.4 eV because of elastic encounters of secondary electrons with the thermal electrons.

The theoretical temperatures in Table 4.10 are subject to several uncertainties. There is no good evidence on the relative abundances of different atoms in the gas, and if C, N, and O atoms tend to be concentrated in grains, the equilibrium temperature may substantially exceed the computed values. Other energy sources, including the inflow of gas to the galactic plane at high velocity, may be significant. Hence the actual temperatures may even exceed those computed for ζ_f equal to ζ_s, with the larger relative increases occurring at the lower values of $n(\text{H I})$.

Values of the cooling time, t_T, defined in equation (4-69) are given in Table 4.11 for different values of the temperature. Since

TABLE 4.11

Cooling Times in H I Regions

	Time t_T in years			
Temperature (°K)	30°	100°	300°	1000°
$n(\text{H I}) = 1 \text{ cm}^{-3}$	1.5×10^5	1.3×10^5	1.4×10^5	2.3×10^5
$n(\text{H I}) = 10 \text{ cm}^{-3}$	2.4×10^4	1.6×10^4	1.6×10^4	2.5×10^4

T exceeds T_E for all entries in the table, Γ_{HR} has been ignored; t_T varies somewhat less rapidly than $1/n_H$ because of the change of n_e/n_H with n_H. Cooling by molecular H_2 has again been ignored If the temperature is increased abruptly by a collision between two clouds, the time required for T to approach T_E will be less than 10^5 years, if n_H exceeds $10\,\text{cm}^{-3}$.

Equipartition of Kinetic Energy

A temperature difference $T_e - T_H$ can result from imperfect equipartition between electrons and H atoms. Let Λ_H be the energy lost by H atoms per cm^3 per sec in inelastic encounters with ions, atoms, and molecules; since the counterbalancing gain appears mostly as electron kinetic energy, in general, then the flow of energy from electrons to H atoms per cm^3 per sec must about equal Λ_H. The flow of energy will actually occur first from the electrons to the positive ions; T_e and T_i are coupled very closely by strong electrostatic interactions. The H atoms will then gain energy from positive ions about 1 order of magnitude more rapidly than from the very light electrons, the larger cross section and the greater momentum exchange for $H - i$ encounters more than offsetting the lower collision frequency.
energy lost by H atoms per cm^3 per sec in inelastic encounters with ions, atoms, and molecules; since the counterbalancing gain

To compute $T_H - T_i$ for a given heat flow, Λ_H, we apply equation (4-11), with heavy positive ions representing the test particles, and hydrogen atoms, the field particles, If we multiply this equation by $3n_i\,3k/2$, we obtain for the heat flow

$$\Lambda_H = \frac{3n_i n(\text{H I})\sigma_s}{m_i}\left(\frac{2kT_H m_H}{\pi}\right)^{1/2} k(T_i - T_H) \qquad (4\text{-}83)$$

where t_s, the slowing-down time of the heavy positive ions moving through the field particles, has been eliminated by use of equation (4-9).

Since m_i occurs in the denominator in equation (4-83), protons are much more effective in transferring kinetic energy to neutral H atoms than are the heavier ions. Since Table 4.10 indicates that the proton particle density is usually greater than the particle

4. INTERACTIONS AMONG INTERSTELLAR PARTICLES 141

density of the heavier ions, we shall consider only the protons here. We ignore a small correction required in equation (4-83) when test and field particles have the same mass. If we take from Table 4.1 the value of σ_s for elastic encounters between H atoms and ions, the variation of this cross section as $1/T_H^{1/2}$ cancels out the $T_H^{1/2}$ in equation (4-83) and we find

$$T_i - T_H = 1.1 \times 10^{24} \frac{\Lambda_H}{n_p n(\text{H I})} \qquad (4\text{-}84)$$

To evaluate $T_i - T_H$ we may set Λ_H equal to Λ_{Hi}, evaluated from equations (4-23) and (4-72) [with $n(\text{H I})$ replacing n_e]. With the cross sections given in Table 4.3 and with numerical values of T and n_p (equal to $n_e - 4.9 \times 10^{-4} n_H$) taken from Table 4.10, one finds that $T_i - T_H$ is at most about $0.1°$ and can generally be ignored [30].

A corresponding analysis for equipartition between charged particles, based on equation (4-12), shows [30] that $T_e - T_i$ is generally less than $T_i - T_H$ by 1 to 2 orders of magnitude in H I regions. Similarly, in H II regions any deviations from equipartition can generally be ignored.

Thermal Instability

If the cooling time, t_T, defined in equation (4-69), is negative, the gas is thermally unstable. In this situation if T exceeds T_E slightly, the kinetic energy will increase, and departures from equilibrium will grow exponentially until t_T changes. If the density, ρ, is constant, t_T will be negative provided that $\Gamma - \Lambda$ increases with increasing T. If the pressure is assumed constant, which seems more realistic, $d\rho/\rho$ will equal $-dT/T$, and there will be a change in $\Gamma - \Lambda$ resulting from the change in n as well as from the change in T. The condition that the total change in $\Gamma - \Lambda$ is negative when T increases gives the following condition [47] for stability

$$T \frac{\partial}{\partial T}(\Gamma - \Lambda) - \rho \frac{\partial}{\partial \rho}(\Gamma - \Lambda) < 0 \text{ for stability} \qquad (4\text{-}85)$$

Since the derivatives are evaluated at the equilibrium temperature, T_E, $\partial(\Gamma - \Lambda)/\partial\rho$ is related to $dT_E/d\rho$, and condition (4-85) may be written as

$$\left[\frac{\partial}{\partial T}(\Gamma - \Lambda)\right]\left[1 + \frac{\rho}{T_E}\frac{dT_E}{d\rho}\right] < 0 \text{ for stability} \qquad (4\text{-}86)$$

Thus thermal instability can appear if either the gains increase more rapidly than the losses, as T is increased above T_E, or if T_E increases more rapidly than $1/\rho$ as the density is decreased. Thermal stability would appear to be present for most conditions in interstellar space, except possibly in regions of relatively low density.

4.5 Equilibrium Properties of the Grains

Interactions between the solid particles of interstellar space, on the one hand, and atoms, molecules and photons, on the other, will affect such physical properties of the grains as the temperature of the solid material in the grains, the electric charge, and the orientation with respect to the magnetic field, **B**. In this section we discuss the steady state produced by such interactions. Evolution of the grains is treated in Section 4.6.

Temperature of the Solid Material

A grain will be heated by absorption of starlight and by collisions with atoms, molecules, and electrons. In a steady state the energy gained in this way will just equal the energy radiated by the grains in the infrared, and the temperature of the solid material within the grain, which we denote by T_s, will adjust itself to satisfy this criterion.

In principle the rates of gain and loss of internal energy are readily computed, though in practice a lack of knowledge as to the properties both of the grains and of the interstellar radiation field makes such computations uncertain. Let H_r and H_a be the heating rate per cm³ due to electromagnetic radiation and atoms, respectively. Then we may write

$$H_r = c n_g \sigma_g \int_0^\infty Q_a(\lambda) U_\lambda \, d\lambda \qquad (4\text{-}87)$$

where c is the velocity of light, n_g is the number of grains per cm³, and σ_g is the geometrical cross section of the grains; $U_\lambda \, d\lambda$ is the energy density of electromagnetic radiation per cm³ between λ and $\lambda + d\lambda$ and $Q_a(\lambda)$ is the efficiency factor for absorption. Similarly, if we neglect any kinetic energy carried off by atoms leaving the grain, H_a is given by

$$H_a = n_g n_a \sigma_g \frac{m_a}{2} \langle w_a^3 \rangle = 4 \left(\frac{2}{\pi m_a} \right)^{1/2} n_g n_a \sigma_g (kT)^{3/2} \qquad (4\text{-}88)$$

where n_a, m_a, and w_a are the particle density, mass, and random velocity of the atoms, and T is the kinetic temperature of the atoms, assumed much greater than T_s. Equation (4-88) assumes a Maxwellian velocity distribution for the atoms; occasional collisions with suprathermal particles, which cause momentary heating of a grain, are discussed in the following section. If we divide equation (4-87) by equation (4-88), and denote by $\bar{Q}_a$ an average value of $Q_a(\lambda)$, the ratio of H_r to H_a becomes

$$\frac{H_r}{H_a} = \left(\frac{\pi m_a}{2} \right)^{1/2} \frac{c \bar{Q}_a U}{4 n_a (kT)^{3/2}} \qquad (4\text{-}89)$$

where U is the total energy density of interstellar radiation, about equal to 7×10^{-13} erg/cm³. If we let T equal 10^4, and consider that the atoms are hydrogen, equation (4-89) yields

$$H_r/H_a = 5.3 \times 10^3 \bar{Q}_a / n_H \qquad (4\text{-}90)$$

If the H atoms are ionized, as is generally the case at 10^4 deg, H_a will be increased by about 1 order of magnitude, since the collision rate between grains and protons will be increased by a factor 3.5 (see below), and the energy lost per collision will also be increased. Moreover, electron–proton recombination on the grain can provide 13.6 eV of energy, increasing the energy gained by about 1 order of magnitude. Since $\bar{Q}_a$ may be as small as 0.1 for dielectric grains, H_r may be less than H_a if the proton density exceeds 10 cm^{-3}. Thus in almost all the observed H II regions, where n_e exceeds this value, the grains may be heated primarily by collisions with protons and electrons. In H I regions, however, where most of the grains are probably located, $T^{3/2}$ is less by 10^3, and H_r exceeds H_a unless n_H exceeds 5×10^5 cm^{-3}.

For such regions of neutral H we may compute the equilibrium temperature as determined by radiative processes only; the total absorption must be balanced by the infrared emission, computed from equation (4-87) on substitution for U_λ its value in thermodynamic equilibrium at the temperature, T_s, of the solid material within the grain. We obtain

$$\int_0^\infty Q_a(\lambda) U_\lambda \, d\lambda = \frac{4\pi}{c} \int_0^\infty Q_a(\lambda) B_\lambda(T_s) \, d\lambda \qquad (4\text{-}91)$$

where $B_\lambda \, d\lambda$ equals $B_\nu \, d\nu$ [see equation (2-4)]. If $Q_a(\lambda)$ is independent of λ, equation (4-91) yields

$$T_s = \left(\frac{U}{a}\right)^{1/4} = 3.1° \qquad (4\text{-}92)$$

where a, the usual radiation constant, is the equilibrium radiation density in ergs/cm^3 at a temperature of 1°. This result is somewhat hypothetical, since in fact $Q_a(\lambda)$ is likely to vary strongly with λ. A somewhat more realistic model is obtained if $Q_a(\lambda)$ is assumed to vary as $1/\lambda$, and the interstellar radiation field is approximated by a Planck function at 10^4 deg, diluted by a factor, W, equal to 10^{-14}; equation (4-91) then gives

$$T_s = 10^{4°} W^{1/5} = 16° \qquad (4\text{-}93)$$

Evidently the value of T_s will depend on the detailed properties of $Q_a(\lambda)$. For a metallic grain $Q_a(\lambda)$ is large in the visible, but small in the infrared, and a temperature between 30° and 100° may be anticipated [51]. Dielectric grains may have relatively

4. INTERACTIONS AMONG INTERSTELLAR PARTICLES

strong absorption bands in the infrared, and in H I regions a value of T_s between 10° and 20°K is not unlikely [51]. As pointed out in Section 3.1, for small $2\pi a/\lambda$, and hence for the infrared wavelengths important in emission, Q_a varies as the particle radius. Hence for particles whose radius, a, exceeds 1μ, for which Q_a averaged over starlight does not depend much on a, the particle temperature decreases with increasing radius; for "dirty ice" grains T_s drops by nearly one-half [52] as the radius increases from 1 to 10μ. Graphite particles are also likely to come to a similarly low temperature [53], but T_s for the hypothetical "free radical" particles is most uncertain. A universal radiation field corresponding to a black-body temperature of about 3°, suggested by the radio measurements and by the excitation of interstellar CN molecules (see Section 4.2), would not have much effect on T_s, since $Q_a(\lambda)$ is not likely to be large at millimeter wavelengths. Evidently the value of T_s cannot be estimated with any precision at the present time. Unless the albedo of the grains is very high, the total energy emitted by the grains should be comparable with the total stellar energy, and infrared measurements should, in principle, provide a determination both of Q_a and of T_s.

Electric Charge

In a steady state a grain cannot continually gain electric charge. Hence if electrons and positive ions stick to the grain or are neutralized with the same probability, the charge on the grain must adjust itself to a sufficiently high negative value so that electrons and positive ions will strike the grain in equal numbers, despite a large difference in their random velocities. We neglect here emission of electrons by the photoelectric effect, a process [54] which within a few parsecs from the brightest early-type stars can produce a positive charge on the grains.

For collisions between charged grains and ions of charge Ze, the effective cross section is πp^2, where p is the collision parameter (distance of closest approach in the absence of forces) for which the ion orbit is just tangent to the surface of the spherical grain.

From conservation of energy and angular momentum we obtain

$$\pi p^2 = \sigma_g \left(1 - \frac{2ZeU}{mw_i^2}\right) \tag{4-94}$$

where U is the electric potential at the surface of the grain in e.s.u., and σ_g is again the geometrical cross section. If all ions had the same random velocity, w_i, the electric charge gained by the grain per second in collision with positive ions would be $n_i w_i Ze\pi p^2$. If U is negative this expression must be integrated over a Maxwellian velocity distribution for all velocities from zero to infinity. For electrons a Maxwellian velocity distribution must again be introduced, but the corresponding expression must now be integrated from a lower limit at which p is zero and the electron kinetic energy is insufficient to overcome the grain potential. Equating the two rates of charge gain which result from these two integrations, assuming that charges of either sign stick to the grain until they are neutralized, yields

$$e^{+eU/kT} = \left(\frac{m_e}{m_i}\right)^{1/2} \left(1 - \frac{ZeU}{kT}\right) \tag{4-95}$$

where we have assumed that only one type of positive ion is present and that n_e equals Zn_i.

In an H II region m_i is the proton mass m_p, $(m_e/m_p)^{1/2}$ equals 1/43, Z is unity, and equation (4-95) has the solution

$$\frac{eU}{kT} = -\frac{Z_g e^2}{akT} = -2.51 \tag{4-96}$$

where $-Z_g e$ is the charge on the grain. Thus the electric charge on the grain increases the number of proton collisions with grains by a factor 3.5. If T is 10,000°, kT is 0.86 eV and the potential of the grain surface is –2.2 V; this corresponds to an excess charge, Z_g, of about 450 electrons, if the grain radius, a, is taken as 3×10^{-5} cm. This electric charge much increases the cross section for deflection of ions in elastic encounters with grains; if equations (4-12) and (4-96) are used to compute t_s, the slowing-down time for charged

4. INTERACTIONS AMONG INTERSTELLAR PARTICLES

grains moving through the gas, the value obtained is several orders of magnitude less than the value computed from equation (4-9) for uncharged grains. In H I regions, where T is only about $100°$, the mean electric charge per grain is only a few electrons; statistical fluctuations of the charge on the grains will be almost as large in these regions as the charge itself. The collisions between neutral H atoms and grains, which are dominant in H I regions, are essentially unaffected by the electric charge on the grains.

Orientation

The polarization of starlight by interstellar grains can be explained only if the grains are optically anisotropic and aligned. Here we shall assume that the grains are effectively spheroidal and discuss how they can be aligned by a magnetic field. The mechanism involved is magnetic relaxation, which produces a dissipation of energy and an associated torque when a grain rotates about an axis at an angle to the magnetic field.

The mechanism of confinement can be explained in terms of particle dynamics [55] or in terms of statistical equilibrium [56]; we follow the statistical approach here. Two effects may be distinguished, one depending on the assumed elongation of the particles, the other depending on the difference of temperature between the gas and the solid material within the grains. The first effect is the tendency of prolate spheroids in kinetic equilibrium each to rotate about an axis perpendicular to the axis of symmetry of the spheroid. If I is the moment of inertia about the symmetry axis, and γI is the corresponding moment about a transverse axis, then for a prolate spheroid γ exceeds unity. When the kinetic energies of rotation are equal about all three axes, as occurs when complete kinetic equilibrium is present, the angular momentum about each axis is proportional to the square root of the moment of inertia about that axis. It follows that the total angular momentum, $\mathbf{J}$, of the spheroid tends to be perpendicular to the axis of symmetry, since the angular momentum about the transverse axis exceeds that about the axis of symmetry by a factor

$\gamma^{1/2}$. Hence in thermodynamic equilibrium prolate spheroids tend to rotate about axes perpendicular to their symmetry axes.

The second effect is the tendency of the rotational energy to be greatest about an axis in space parallel to **B**, if the gas temperature, T, exceeds T_s, the temperature of the solid material composing the grain. Rotation about this axis is affected only by collisions with gas atoms, and the rotational energy will tend to have its equipartition value at the temperature T. Rotation about the axes perpendicular to **B**, however, is affected by the torque associated with magnetic relaxation; the mean torque will tend to slow down the rotation of the particle, while spontaneous fluctuations of magnetization will tend to accelerate the rotation. If no collisions were present, the rotational energy would tend to have its equipartition value, $\frac{1}{2}kT_s$, corresponding to the temperature, T_s, since this temperature determines the random accelerating torques resulting from thermal fluctuations of magnetization. When magnetic torques and collisions with atoms are both present, the mean kinetic rotational energy about each axis perpendicular to **B** will tend to have some average value, $\frac{1}{2}kT_{av}$, intermediate between $\frac{1}{2}kT$ and $\frac{1}{2}kT_s$. If T_{av} is substantially less than T, then the total angular momentum will tend to be oriented parallel to **B**. As a result of these two effects, elongated particles will tend to rotate about **B** with their major axes perpendicular to **B**.

A detailed analysis must take into account the complex motion of a freely rotating spheroid, whose instantaneous angular velocity vector is not fixed in space. However, the total angular momentum, **J**, of a spheroid is, of course, constant in the absence of external torques. Let β be the angle between **J** and **B**, and θ, the angle between **J** and the principal axis of the spheroid. Magnetic torques and collisions with atoms will change both θ and β, and a precise theory must discuss the distribution of the two angles θ and β, over their possible values. The factors determining the distribution of θ and β in the general case are too complex for a complete analysis, but an approximate model may be provided by the assumption that the distributions of θ and β are independent, with $f_\theta(\theta)d\theta$ and $f_\beta(\beta)d\beta$ the fractions of grains with θ and β

4. INTERACTIONS AMONG INTERSTELLAR PARTICLES

lying within the ranges $d\theta$ and $d\beta$, respectively. Consistently with this assumption of a separable distribution function, f_θ is evaluated with neglect of the magnetic field, while f_β is evaluated with the elongation of the particles ignored.

The distribution function $f_\theta(\theta)$ for prolate spheroids in thermal equilibrium will be considered first. To evaluate this function we consider the distribution of angular momenta J_A and J_B, where J_A is about the axis of symmetry, while J_B is in a plane perpendicular to this axis. Let $f_A(J_A)dJ_A$ be the fraction of grains for which J_A lies within the interval dJ_A. Then in kinetic equilibrium at a temperature T we have the usual Maxwell-Boltzmann distribution

$$f_A(J_A) = \frac{1}{(2\pi IkT)^{1/2}} e^{-J_A^2/2IkT} \qquad (4\text{-}97)$$

The function f_B is similar, but with the corresponding moment of inertia γI replacing I; in addition a factor $2\pi J_B$ appears in the numerator, since all directions perpendicular to the axis of symmetry are included, and for normalization an additional factor $(2\pi\gamma IkT)^{1/2}$ appears in the denominator. We may now write $f_\theta(\theta)\,d\theta$ as the integral

$$f_\theta(\theta)\,d\theta = 2\pi \iint \frac{J_B\,dJ_B\,dJ_A}{\gamma(2\pi IkT)^{3/2}} \exp\left\{-\frac{J_A^2}{2IkT} - \frac{J_B^2}{2\gamma IkT}\right\} \qquad (4\text{-}98)$$

integrated over all values of J_B and J_A such that

$$\theta - \frac{d\theta}{2} < \tan^{-1}\frac{J_B}{J_A} < \theta + \frac{d\theta}{2} \qquad (4\text{-}99)$$

To satisfy equation (4-99) we make the substitution

$$J_B = J_A \tan\theta; \quad dJ_B = J_A \sec^2\theta\,d\theta \qquad (4\text{-}100)$$

and then integrate equation (4-98) over J_A from $-\infty$ to $+\infty$, obtaining

$$f_\theta(\theta) = \frac{\gamma^{1/2}}{2} \frac{\sin\theta}{(\gamma\cos^2\theta + \sin^2\theta)^{3/2}} \qquad (4\text{-}101)$$

The temperature T has disappeared from f_θ, and the distribution depends only on γ. When γ is large, $f_\theta/2\pi \sin\theta$, which gives the distribution of θ per unit solid angle, has a sharp peak at θ equal to $\pi/2$, and prolate spheroids tend to rotate about axes perpendicular to the axis of symmetry.

Next we consider the distribution function f_β. As pointed out above, the grains will be assumed spherical in the evaluation of this function. Again we consider angular momenta about two axes, this time letting J_z be the angular momentum parallel to **B**, while J_y represents the angular momentum in a plane perpendicular to **B**. The distribution function $f_z(J_z)$ is identical with $f_A(J_A)$ in equation (4-97), with T equal to the gas temperature. However, $f_y(J_y)$ differs from $f_B(J_B)$; the moment of inertia is now equal to I, since the grains are assumed spherical, and the rotational temperature differs from T, since magnetic torques affect the rotation about this axis. If collisions with the gas are much less important than these magnetic effects, then the distribution of J_y will correspond to thermodynamic equilibrium at the temperature T_s; i.e., T_s will replace T in going from $f_B(J_B)$ to $f_y(J_y)$. If collisions dominate, T_s drops out and the distribution of J_y corresponds to equilibrium at the gas temperature, T. In the more general case, where both gas collisions and magnetic torques are important, it may be shown [56] that the distribution of J_y is still given by the Maxwell-Boltzmann formula at a temperature T_{av} intermediate between T and T_s, given by

$$T_{av} = \frac{T + \delta T_s}{1 + \delta} \qquad (4\text{-}102)$$

The quantity δ is the ratio between the mean magnetic torque and the mean torque to collisions between grains and atoms, which we take to be neutral hydrogen. For a particle of volume V rotating about an axis perpendicular to **B**, the former torque is [57] $VB^2\chi''$, where χ'' is the imaginary part of the magnetic susceptibility. The latter torque may be evaluated by straightforward computations [56], if it is assumed that the atoms stick

4. INTERACTIONS AMONG INTERSTELLAR PARTICLES

and then come off the grain isotropically relative to the moving surface. One obtains the relation

$$\delta = \frac{B^2}{an_H} \times \frac{\chi''}{\omega} \times \left(\frac{\pi}{2m_H kT}\right)^{1/2} \qquad (4\text{-}103)$$

The function f_β may be found from f_z and f_y in exactly the same manner as f_θ was found from f_A and f_B, giving

$$f_\beta(\beta) = \frac{1}{2}\left(\frac{T_{av}}{T}\right)^{1/2} \frac{\sin\beta}{\left(\frac{T_{av}}{T}\cos^2 + \sin^2\beta\right)^{3/2}} \qquad (4\text{-}104)$$

If T_{av} is appreciably less than T, $f_\beta/2\pi \sin\beta$ will have a sharp peak at β equal to zero and particles will tend to rotate with **J** parallel (or antiparallel) to **B**; in this direction magnetic torques are inoperative and the effective rotational temperature is greater than that for rotation about the transverse axes.

From a rigorous standpoint this analysis shows only that spheres in a magnetic field tend to rotate about axes parallel to **B**, while prolate spheroids, in the absence of a magnetic field, tend to rotate with their axes of symmetry perpendicular to the axis of rotation. It seems reasonable to assume that for actual spheroids in a magnetic field the distribution function $f_\theta(\theta)f_\beta(\beta)$ determined in this way provides a reasonable first approximation to the actual distribution, with the spheroids tending to rotate about axes parallel to **B** and perpendicular to the symmetry axes. This model provides a quantitative if approximate basis for interpretating the measures of interstellar polarization.

We have seen in Chapter 3 that rather good orientation of dielectric spheroids is required to explain the polarization observations. Elongated grains with moderately large γ and with T_s appreciably less than the gas temperature seem not unlikely, although no independent confirmatory evidence is available. The value of δ, which may be computed from equation (4-103), must be at least unity for appreciable orientation. For paramagnetic relaxation χ''/ω is about $3 \times 10^{-12}/T_s$ in dielectric materials at a temperature T_s. If we let T and T_s equal 100° and 10°K, respectively, and set a equal to 3×10^{-5} cm, we find from equation (4-103)

$$\delta = 8.3 \times 10^{10} (B^2/n_H) \qquad (4\text{-}105)$$

If n_H equals 10/cm³, a typical value in dust clouds, δ will equal unity only if B equals 1.1×10^{-5} G, a rather high value. If graphite particles or complex grains with graphite cores are assumed, the magnetic field required for orientation is about the same [52], provided the grains have the large dimensions required by the dependence of polarization on wavelength (see Section 3.3). However, if the iron atoms, which are primarily responsible for the magnetic effects in paramagnetic grains, are gathered together in little clumps, χ''/ω can be enhanced [56] by several orders of magnitude, with a large reduction in the magnetic field strength required for orientation. Similarly, if the iron atoms, possibly in tightly bound molecules with oxygen and magnesium, form one or more separate ferromagnetic regions within the grains, strong orientation is possible [56] even if B is much less than 10^{-6} G. In the absence of firm evidence on the magnitude of **B** or on the properties of grains one can conclude only that good orientation of interstellar grains by a magnetic field is not inconsistent with present knowledge.

4.6 Evolution of Grains and Formation of Molecules

Collisions between atoms and grains in the interstellar gas can produce a gradual evolution both of the grains and the gas. The distribution of grain sizes is presumably the result of this evolution. So little is known about the grains that a definitive discussion of the relevant physical and chemical processes is impossible at present. Some of the processes that might occur will be described here, and the relevant time scales computed.

Growth of Grains

One of the most significant results to be anticipated from the collisions between atoms and grains is the growth of the grains

4. INTERACTIONS AMONG INTERSTELLAR PARTICLES

themselves. Since most of the obscuring clouds are likely to be H I regions, we consider collisions between neutral atoms, of mass Am_H, and grains of radius a and particle density n_g. For any one atom the mean time, t_{ag}, between collisions with a grain, is given by

$$t_{ag} = \frac{1}{w_a n_g \sigma_g} = 6.3 \times 10^8 F_c A^{1/2} \text{ years} \quad (4\text{-}106)$$

where we have replaced $n_g \sigma_g$ by its mean value of 3.2×10^{-22} cm²/cm³, found in Section 3.2, divided by the concentration factor F_c, and where w_a has been replaced by its r.m.s. value at 100°K. If we assume that F_c equals 0.07 (see Table 3.4) then for oxygen atoms

$$t_{ag} = 1.8 \times 10^8 \text{ years} \quad (4\text{-}107)$$

While the time t_{ag} computed from equation (4-106) is not likely to be seriously in error for H I regions, the effects produced by collisions are much more uncertain. If the sticking probability for atoms is denoted by ξ_a, then accretion of atoms will increase the mass of the grains at a rate given by

$$\frac{dm}{dt} = \pi a^2 w_a n_a A m_H \xi_a \quad (4\text{-}108)$$

giving an increase of grain radius, a, at the rate

$$\frac{da}{dt} = \frac{w_a \rho_a \xi_a}{4\rho_s} = \frac{3.9 \times 10^4 \rho_a \xi_a}{A^{1/2} \rho_s} \quad (4\text{-}109)$$

where ρ_a is the density of the interstellar atoms which adhere to the grain, equal to $n_a A m_H$, and ρ_s is the density of solid matter inside the grain. Thus if ρ_s equals 1 g/cm³, a typical value for dielectric grains, and $\rho_a/A^{1/2}$ equals 1.0×10^{-25} g/cm³, about the value obtained by summing $\rho_a/A^{1/2}$ for the elements heavier than H in Table 4.7 (excluding the noble gases) for a hydrogen particle density of ten atoms/cm³, we find

$$da/dt = 1.2 \times 10^{-13} \xi_a \text{ cm/year} \quad (4\text{-}110)$$

Even if ξ_a is only about 0.1, the radius of a grain will grow to about the observed value in some 10^9 years. For H_2 and He it seems fairly clear that ξ_a is very small, since the rate of evaporation of these substances is high [58] at a temperature of 5°K or more. For heavier elements van der Waals binding seems not unlikely, and possibly even chemical binding, though some activation energy might be required in this latter case. It is generally assumed that ξ_a is sufficiently large so that the grains have grown to their present radii in accordance with equation (4-110). The origin of the original condensation nuclei is obscure. Nuclei of graphite may form in the atmospheres of the cooler carbon stars and be expelled by radiation pressure [53].

Disruption of Grains

If there were no process tending to disrupt the grains, the heavy atoms should mostly be bound within these solid particles, at least in the denser clouds. The data do not exclude this possibility. For the composition in Table 4.7 about 1.8% by mass of the interstellar material is in the form of solid particles, if C, N, and O are assumed chemically bound in stable hydride molecules within the grains. We have seen in Section 3.4 that the density of the grains is about 1% of the gas density, but this estimate is clearly uncertain within a factor 2. However, if most of the heavy atoms were in solid particles, the shape of the distribution function for grain radii might be expected to vary more than is observed from one cloud to another, depending on the number of condensation nuclei, the rate of growth, etc. The approximate uniformity of the observed reddening curves suggests that an equilibrium between formation, growth, and disruption may be present, with the reaction rates depending on density but not the final equilibrium.

Collisions between clouds at a relative velocity of some 10 to 20 km/sec can disrupt grains. Such collisions are treated in Section 5.1 below. Here we are not concerned with the complex behavior of the gas but rather with the collisions between the grains in

4. INTERACTIONS AMONG INTERSTELLAR PARTICLES

each of the two colliding clouds. While the grains in each of the two clouds will tend to be slowed down by the surrounding gas, some of the grains will collide at speeds greater than the few km/sec required [59] for complete evaporation. To compute the probability of such collisions we can compare λ_{gg}, the mean free path for a collision between two grains, and λ_{gH}, the mean free path for slowing down of a grain by collisions with H atoms. Since the collision cross section between two identical grains is four times the geometrical cross section of each one separately, we may multiply by 4 the value 3.2×10^{-22} cm^{-1} found for $n_g \sigma_g$ (see Section 3.2) and obtain a value of $7.8 \times 10^{20} F_c$ cm for λ_{gg}. The value of λ_{gH} is the distance over which the mass of hydrogen and helium intercepted by the grain equals the mass of the grain. Hence

$$\lambda_{gH} = \frac{4a\rho_s}{3\rho_H(1 + 4n_{He}/n_H)} \qquad (4\text{-}111)$$

and for a equal to 3×10^{-5} cm, ρ_s equal to 1 g/cm^3, n_{He}/n_H equal to 0.1, and ρ_H equal to $1/F_c$ times its average value, 2×10^{-24} g/cm^3, we obtain

$$\lambda_{gH}/\lambda_{gg} = 0.018 \qquad (4\text{-}112)$$

This value provides an upper limit on the probability of evaporation per cloud collision. Grazing collisions between grains will not produce complete evaporation; the compression of the gas by shock waves between the two colliding clouds will decrease λ_{gH}, especially if the cooling rate is rapid, without affecting λ_{gg}. Evidently at most 1% probability of evaporation per collision may be anticipated. If 10^7 years is taken as the time between cloud collisions, 10^9 years becomes the disruption time of the grains. A detailed calculation [59] of $f(r)$, the distribution function of grain radii under these conditions, gives a result that appears to be consistent with most of the observations of selective extinction, but not with preliminary measures between 2000 and 3000 Å.

A more gradual process of disruption is provided by sputtering, a process by which individual atoms are knocked out of grains. At grazing incidence a proton energy of a few electron volts is sufficient to knock water molecules, for example, out of a grain. Heavier incident atoms transfer more energy in an elastic collision and helium atoms of a few eV can produce sputtering even at normal incidence [60]. If grains spend an appreciable length of time in H II regions, this process can probably diminish the mass of each particle appreciably.

Disruption of very small grains can, in principle, be produced by low-energy cosmic rays. When an energetic nucleus of charge Ze and velocity βc passes centrally through a spherical grain of radius a and density ρ_s, the energy ΔE lost by electron excitation is given by [23]

$$\Delta E = 2a \times \frac{4\pi n_e e^4 Z^2}{mc^2 \beta^2} \ln\left\{\frac{2mc^2\beta^2}{\bar{E}(1-\beta^2)} - \beta^2\right\}$$
$$= 7.0 \times 10^5 \frac{aZ^2 \rho_s}{\beta^2} \left\{4.53 + \log \frac{\beta^2}{1-\beta^2} - 0.43\,\beta^2\right\} \text{ eV} \quad (4\text{-}113)$$

n_e is the number of electrons per cm^3 within the grain, which we have set equal to $\rho_s/2m_H$, m is the electron mass, and $\bar{E}$ is a mean ionization energy, which we have taken to be 30 eV. Thus a 2 MeV proton, with β equal to 1/15, which passes through a grain of radius 10^{-7} cm, containing about 100 molecules of H_2O, heats up the grain by about 35 eV, providing about the 0.2–0.4 eV per molecule required for complete evaporation. If P_{ev} represents the probability per unit time of evaporation in this way, the ratio of P_{ev} to ζ_f, the corresponding ionization rate for H atoms, equals the ratio of the grain cross section, πa^2, to $5\sigma_{HR}/3$, where σ_{HR} is the ionization cross section, and the factor 5/3 is the ratio of the total number of ion pairs produced to the number of primary ion pairs. If we assume that protons from supernova shells increase ζ_f up to the value ζ_s, then as a first approximation we may evaluate σ_{HR} for protons of energy 2 MeV. In this way we find from equation (4-29) for a grain of radius 10^{-7} cm

$$P_{ev} = 6 \times 10^{-5} \text{ year}^{-1} \quad (4\text{-}114)$$

4. INTERACTIONS AMONG INTERSTELLAR PARTICLES 157

Comparison of equations (4-110) and (4-114) indicates that a grain of this small size will evaporate before it can grow appreciably, unless $\xi_a n_H$ exceeds unity. Large grains, of the size observed, will not be much affected by the passage of an energetic particle, since the mean temperature will not be increased very much. Evidently if energetic particles increase ζ_f to its upper limit, the interstellar grains are not likely to grow slowly from interstellar condensation nuclei, and a stellar origin of the sort referred to above would seem indicated.

Molecule Formation

Collisions between atoms and grains can also lead to the formation of molecules, the grains serving as a catalyst for this reaction. If half of the collisions between atoms and grains led to the escape of free molecules into the gas, the gas in a typical cloud would become composed of molecules in a time comparable with t_{ag}, or about 3×10^8 years for oxygen. Some molecules will subsequently be dissociated by interstellar radiation either directly or indirectly through ionization followed by dissociative recombination (see Section 4.3). A detailed analysis [45] of such processes in the case of the CH molecule indicates that despite large uncertainties in the rate coefficients, the observations cannot be explained unless an appreciable fraction of the C^+ ions striking the grains evaporate as neutral CH. Molecule formation by this process is not much more rapid than by radiative association of C^+ and an H atom to form CH^+. However, the formation of neutral molecules from C^+ ions on the surface of grains helps to explain the observed near equality of $n(CH)$ and $n(CH^+)$ [see equation (2-64)].

The CH^+ lines observed in the spectra of unreddened stars of types B2–B8 may be attributed to evaporation of methane from the grains, followed by gradual dissociation and ionization. Since a grain must come within roughly a parsec from a B8 star to reach the temperature of about 30°K required for evaporation of CH_4, this process is probably not very important for the interstellar medium as a whole. However, the observed value of

$N(\mathrm{CH}^+)$ in the spectra of these stars is consistent [45] with what one would expect from such a process, as are also the radial velocities of the CH^+ molecules, which are approaching the stars in almost all cases. On the other hand, the observed CH line at $\lambda 4300$ in the spectra of several unreddened stars of types B2–B5 is too strong to be explained in this way, since CH should be ionized rapidly within a parsec from a B star.

Reactions on the surface of grains may be of particular importance for the formation of molecular hydrogen, since as pointed out in Section 4.3 radiative association is forbidden for this homonuclear molecule. During the time interval, t_{ag}, equal to about 4×10^7 years for H atoms, much of the interstellar hydrogen may pass through a molecular phase, if most collisions between H atoms and grains lead to H_2 formation. The dissociation time of 600 years quoted in Section 4.3 then gives a value of about 10^{-5} for the ratio $n(H_2)/n_H$. While theoretical arguments are somewhat conflicting [61, 62], it is possible that the formation of H_2 on grains requires activation energy. If so, molecular hydrogen will form only during collisions between clouds, and the mean ratio of $n(H_2)$ to n_H may be several orders of magnitude [44] less than 10^{-5}. In a relatively dense cloud, on the other hand, the radiation in the Lyman bands of H_2 may be entirely absorbed, dissociation may be absent, and an appreciable fraction of the hydrogen may be molecular.

References

1. S. Chandrasekhar, *Principles of Stellar Dynamics*, University of Chicago Press, 1942, Chapter 2 and Section 5.6.
2. A. Dalgarno and A. C. Allison, informal communication.
3. J. O. Hirschfelder, C. F. Curtiss and R. B. Bird, *Molecular Theory of Gases and Liquids*, Wiley, New York, 1954, Section 8.4.
4. D. E. Osterbrock, *Astrophys. J.*, **134**, 270 (1961).
5. A. Dalgarno, *Phil. Trans. Roy. Soc.* (*London*), Ser. A, **250**, 426 (1958).
6. C. Schwartz, *Phys. Rev.*, **124**, 1468 (1961).
7. L. Spitzer, *Physics of Fully Ionized Gases*, Interscience, New York, 2nd rev. ed., 1962, Chapter 5.
8. M. J. Seaton, *Rev. Mod. Phys.*, **30**, 979 (1958), (*Intern. Astron. Union Symp.* No. 8).

4. INTERACTIONS AMONG INTERSTELLAR PARTICLES

9. S. J. Czyzak, in *Stars and Stellar Systems*, Vol. 7, Univ. of Chicago Press, Chicago, 1968, p. 403.
10. M. J. Seaton, *Atomic and Molecular Processes*, Academic Press, New York 1962, p. 374.
11. S. J. Czyzak, T. J. Kruger, P. de A. P. Martins, H. E. Saraph, M. J. Seaton and J. Shemming, Intern. Astron. Union Symp. No. 34, 1967 in press.
12. D. E. Osterbrock, *Astrophys. J.*, **142**, 1423 (1965).
13. R. H. Garstang, *Monthly Notices Roy. Astron, Soc.*, **111**, 115 (1951).
14. M. J. Seaton, *Ann. d'Astrophys.*, **18**, 188 (1955).
15. F. J. Smith, *Planetary Space Sci.*, **14**, 929 (1966).
16. A. Dalgarno and M. R. H. Rudge, *Astrophys. J.*, **140**, 800 (1964).
17. F. J. Smith, *Planetary Space Sci.*, **14**, 937 (1966).
18. A. C. Allison and A. Dalgarno, *Proc. Phys. Soc. (London)*, **90**, 609 (1967).
19. G. B. Field, W. B. Somerville and K. Dressler, *Ann. Rev. Astron. Astrophys.*, **4**, 207 (1966).
20. W. J. Karzas and R. Latter, *Astrophys. J., Suppl.*, **6**, 167 (1961).
21. A. Burgess, *Astrophys. J.*, **139**, 776 (1964).
22. D. R. Bates and A. Dalgarno, *Atomic and Molecular Processes*, Academic Press, New York, 1962, p. 243.
23. H. Bethe, *Handbuch der Physik*, Springer, Berlin, 1933, Vol. **24**, Part I, pp. 518–520.
24. A. Dalgarno and G. W. Griffing, *Proc. Roy. Soc. (London), Ser. A*, **248**, 415 (1958).
25. L. H. Aller and W. Liller, in *Stars and Stellar Systems*, Vol. 7, Univ. of Chicago Press, Chicago, 1968, p. 483.
26. L. H. Aller and D. Bohm, *Astrophys. J.*, **105**, 131 (1947).
27. G. B. Field and J. L. Hitchcock, *Astrophys. J.*, **146**, 1 (1966).
28. R. H. Dicke, P. J. E. Peebles, P. G. Roll and D. T. Wilkinson, *Astrophys. J.*, **142**, 414 (1965).
29. G. B. Field, *Astrophys. J.*, **129**, 551 (1959).
30. L. Spitzer and M. G. Tomasko, *Astrophys. J.*, **152**, 971 (1968).
31. B. Strömgren, *Astrophys. J.*, **89**, 526 (1939).
32. R. J. Gould, T. Gold and E. E. Salpeter, *Astrophys. J.*, **138**, 408 (1963).
33. M. J. Seaton, *Monthly Notices Roy. Astron. Soc.*, **119**, 81 (1959).
34. L. Spitzer, in *Stars and Stellar Systems*, Vol. 7, Univ. of Chicago Press, Chicago, 1968, p. 1.
35. K. S. Krishna Swamy and C. R. O'Dell, *Astrophys. J.*, **147**, 529 (1967).
36. L. Spitzer and J. L. Greenstein, *Astrophys. J.*, **114**, 407 (1951).
37. D. E. Osterbrock, *Astrophys. J.*, **135**, 195 (1962).
38. C. R. O'Dell, *Astrophys. J.*, **142**, 1093 (1965).
39. D. G. Hummer and M. J. Seaton, *Monthly Notices Roy. Astron. Soc.*, **127**, 217 (1964).
40. M. J. Seaton, *Monthly Notices Roy. Astron. Soc.*, **111**, 368 (1951).

41. A. Weigert, *Wiss. Zeit. Friedrich-Schiller Univ., Jena, Math.–Naturwiss. Reihe*, **4**, 435 (1955).
42. L. H. Aller, *Abundances of the Elements*, Interscience, New York, 1961.
43. D. C. Morton, *Astrophys. J.*, **151**, 285 (1968).
44. T. P. Stecher and D. A. Williams, *Astrophys. J.*, **149**, L29 (1967).
45. D. R. Bates and L. Spitzer, *Astrophys. J.*, **113**, 441 (1951).
46. T. O. Carroll and E. E. Salpeter, *Astrophys. J.*, **143**, 609 (1966).
47. G. B. Field, *Astrophys. J.*, **142**, 531 (1965).
48. R. M. Hjellming, *Astrophys. J.*, **143**, 420 (1966).
49. K. Takayanagi and S. Nishimura, *Publ. Astron. Soc. Japan*, **12**, 77 (1960).
50. L. Spitzer, *Astrophys. J.*, **109**, 337 (1949).
51. H. C. van de Hulst. *Rech. Astron. Obs. Utrecht*, **11**, Part 1 (1946).
52. J. M. Greenberg, in *Stars and Stellar Systems*, Vol. 7, Univ. of Chicago Press, Chicago, 1968, p. 221.
53. N. C. Wickramsinghe, *Interstellar Grains*, Chapman and Hall, London, 1967.
54. W. G. Mathews, *Astrophys. J.*, **147**, 965 (1967).
55. L. Davis and J. L. Greenstein, *Astrophys. J.*, **114**, 206 (1951).
56. R. V. Jones and L. Spitzer, *Astrophys. J.*, **147**, 943 (1967).
57. L. Davis, *Astrophys. J.*, **128**, 508 (1958).
58. H. C. van de Hulst, *Rech. Astron. Obs. Utrecht*, **11**, Part 2 (1949).
59. J. H. Oort and H. C. van de Hulst, *B.A.N.*, **10**, 187 (1946), No. 376.
60. E. Langberg, *Phys. Rev.*, **111**, 91 (1958).
61. R. J. Gould and E. E. Salpeter, *Astrophys. J.*, **138**, 393 (1963).
62. H. F. P. Knaap, C. J. N. van den Meydenberg, J. J. M. Beenakker and H. C. van de Hulst, *Interstellar Grains* (Symposium at Troy, N.Y., 1965), U.S. Govt. Printing Office Washington D.C., NASA SP–140, 1967, p. 253.

Chapter 5

Dynamics of the Interstellar Gas

5.1 Dynamical Principles and Problems

Earlier sections have shown that the interstellar medium comprises a gas of atoms, ions, molecules, and grains, all in kinetic equilibrium, together with an assembly of energetic particles, gyrating in a magnetic field. The density and temperature of the gas are known to vary by orders of magnitude from one location to another, with resultant variations in the state of ionization and dissociation. The dynamical interactions occurring in this inhomogeneous and anisotropic magnetofluid are enormously complicated, with many types of processes possible [1, 2]. Observational contact with these phenomena is in many cases slight, and the discussion in this Chapter will be limited to a few dominant processes that are likely to produce major effects in the interstellar medium.

This first section deals with the basic dynamical equations and their applications, together with a brief discussion of the energy balance in the interstellar cloud motions. Subsequent sections in this chapter analyze the equilibrium density distribution of the gas, and several mechanisms that can accelerate this interstellar material to the velocities observed.

Basic Equations

The change of **v**, the fluid velocity of the gas, is determined by the usual momentum equation, which we write in the form

$$\rho \frac{D\mathbf{v}}{Dt} = \rho \left(\frac{\partial \mathbf{v}}{\partial t} + \mathbf{v} \cdot \nabla \mathbf{v} \right)$$
$$= -\nabla p - \frac{1}{8\pi} \nabla B^2 + \frac{1}{4\pi} \mathbf{B} \cdot \nabla \mathbf{B} - \rho \nabla \phi \quad (5\text{-}1)$$

In this equation D/Dt denotes the Lagrangian time derivative, following a fluid element, while $\partial/\partial t$ represents the Eulerian time derivative at a fixed location. The quantities ρ and ϕ, which denote the density and the gravitational potential, respectively, are determined by the equation of continuity and Poisson's equation

$$\frac{D\rho}{Dt} = \frac{\partial \rho}{\partial t} + \mathbf{v} \cdot \nabla \rho = -\rho \nabla \cdot \mathbf{v} \quad (5\text{-}2)$$

$$\nabla^2 \phi = 4\pi G \rho \quad (5\text{-}3)$$

The change in time of the magnetic field, **B**, is determined by

$$\frac{\partial \mathbf{B}}{\partial t} = \nabla \times (\mathbf{v} \times \mathbf{B}) + \frac{\eta}{4\pi} \nabla^2 \mathbf{B} \quad (5\text{-}4)$$

subject to the usual condition

$$\nabla \cdot \mathbf{B} = 0 \quad (5\text{-}5)$$

Equation (5-4), in which η represents the electrical resistivity in electromagnetic units, is only approximate [2], but is generally valid under most interstellar conditions. The pressure p includes both the pressure, p_G, of the gas and the pressure, p_R, of the energetic particles or cosmic rays. The gas pressure, p_G, is given by the usual perfect gas law,

$$p_G = \frac{\rho k T}{\mu} = \rho C^2 \quad (5\text{-}6)$$

5. DYNAMICS OF THE INTERSTELLAR GAS

where μ is the mean mass per particle, in grams, k is the Boltzmann constant, and the temperature T is determined by the energy equation (4-68). Turbulence in the gas may be taken approximately into account by including a turbulent pressure in p_G, as in equation (5-33). The gradient of cosmic-ray pressure perpendicular to **B** enters into equation (5-1) in the same way as ∇p_G; however, if p_R has a gradient parallel to **B**, the pressure is anisotropic, and a more complex analysis is required.

Equation (5-1) is sometimes called a "macroscopic equation," since it describes the change of the macroscopic fluid velocity, **v**, defined as the average of the particle velocity, **w**, for all the particles within a small volume element. One can also write down the "microscopic equations," which govern the motion of an individual particle in magnetic and electric fields [2]. The most important result of the microscopic equations is the well-known helical motion of a charged particle around the magnetic lines of force. If the particle mass and charge are m and Ze, respectively, where e is the charge on the proton in e.s.u., then the angular frequency, ω_B, of this helical motion, called the "gyration frequency," or "cyclotron frequency," is

$$\omega_B = \frac{ZeB}{mc} \tag{5-7}$$

The radius, a, of the helix, called the "radius of gyration," is given by

$$a = w_\perp / \omega_B \tag{5-8}$$

where $w_\perp$ is the component of the particle velocity perpendicular to **B**.

The macroscopic equations above have a number of simple consequences [2, 3]. In the absence of fluid velocities the time required for the magnetic field to change by ohmic dissipation is of order $4\pi L^2/\eta$, where L is the scale size of the magnetic field. Even for L as small as 1 A.U. this time will exceed 10^9 years for a typical interstellar conductivity, and for larger scales the field is effectively constant during the age of the Universe. In the presence of material motions equation (5-4) implies that the magnetic flux through any circuit moving with the fluid is constant, provided η is ignored. Hence the lines of magnetic force may be regarded as following the fluid in its motion and are said to be "frozen" into the fluid.

For smaller disturbances about an equilibrium condition the equations above may be replaced by a somewhat simpler set of linear equations for the disturbances (see Section 6.1). These linearized equations permit a number of well-known types of wave motions [2]. If **B** and ϕ are ignored and ρ is assumed nearly uniform, pressure disturbances propagate at the sound velocity. If we consider disturbances with a period long compared to the cooling time, t_T, discussed in Section 4.4, the temperature will remain equal to T_E, its value in radiative equilibrium, and the disturbance will propagate at the sound speed, V_S, given by

$$V_S{}^2 = \frac{kT_E}{\mu}\left\{1 + \frac{\rho}{T_E}\frac{dT_E}{d\rho}\right\} \tag{5-9}$$

We shall usually ignore the correction factor in parentheses, and set V_S equal to the isothermal sound speed, C, at the temperature T_E [see equation (5-6)]. If a uniform magnetic field is present, equations (5-1) and (5-4) may be combined to show that disturbances transverse to **B** will travel along **B** at the Alfvén speed, V_A, given by

$$V_A{}^2 = \frac{B^2}{4\pi\rho} \tag{5-10}$$

These transverse waves traveling along the field lines are similar to a transverse wave traveling along a string under tension.

Virial Theorem

To derive this useful theorem, which applies to an entire system within some closed surface S, one takes the scalar product of the position vector, **r**, with equation (5-1) and integrates over the volume interior to S. On integrating by parts one obtains

5. DYNAMICS OF THE INTERSTELLAR GAS

$$\frac{1}{2}\frac{D^2 I}{Dt^2} = 2T + 3U + \mathcal{M} + W$$

$$+ \frac{1}{4\pi}\int_S (\mathbf{r}\cdot\mathbf{B})\mathbf{B}\cdot d\mathbf{S} - \int_S \left(p + \frac{B^2}{8\pi}\right)\mathbf{r}\cdot d\mathbf{S} \qquad (5\text{-}11)$$

where the quantities I, T, U, $\mathcal{M}$, and W are defined by the volume integrals over the system within the surface S,

$$I = \int \rho r^2 \, dV \qquad T = \frac{1}{2}\int \rho v^2 \, dV \qquad U = \int p \, dV$$

$$\mathcal{M} = \frac{1}{8\pi}\int B^2 \, dV \qquad W = -\int \rho \mathbf{r}\cdot\nabla\phi \, dV \qquad (5\text{-}12)$$

The quantity I is a generalized moment of inertia. The remaining integrals are all energies; T is the kinetic energy of the fluid, U equals two-thirds of the random kinetic energy of the thermal particles in the gas plus one-third of the energy of the relativistic particles, and $\mathcal{M}$ is the magnetic energy within the surface S.

The integral W is the total gravitational energy of the system only if any masses outside the surface S can be ignored in the computation of the potential. This result may be seen if W is written in the form

$$W = -\sum_j m_j \mathbf{r}_j \cdot \sum_k \frac{Gm_k(\mathbf{r}_j - \mathbf{r}_k)}{|(\mathbf{r}_j - \mathbf{r}_k)|^3} \qquad (5\text{-}13)$$

where j is summed over all masses within the system, while k is summed over all gravitating masses. If masses outside the system can be ignored, then each interaction will be counted twice in the double sum, with r_j and r_k interchanged. As a result we have in this case

$$W = -\sum_{j<k} \frac{Gm_j m_k}{|(r_j - r_k)|} \qquad (5\text{-}14)$$

which equals the total gravitational energy of the system.

Equation (5-11) may be applied to some components of a system, as, for example, the interstellar gas, magnetic field, and relativistic particles, but excluding the stars. In this case W is not generally equal to the gravitational energy of the components considered.

Shock Fronts

When a pulse of increased pressure, with an appreciable amplitude, propagates through a gas, the front of the pulse tends to steepen, because the sound velocity is higher in the compressed region; this steepening progresses until a nearly discontinuous shock front is formed. Such shock fronts generally appear whenever supersonic motions are present, and must therefore be expected in the interstellar gas, where the cloud velocities are generally much greater than the sound speed. We consider here the properties of such a front, idealized as a one-dimensional disturbance propagating through a homogeneous medium with a constant velocity, u_1. Let the direction of motion be taken as the x axis; all quantities are then functions of x and t only.

The situation is more conveniently analyzed in a frame of reference traveling with the velocity u_1; in this frame the flow is steady, and all quantities are functions of x only. The undisturbed material enters the shock at the plane x_1, where its velocity, density, and temperature are u_1, ρ_1, and T_1, respectively. At the other side of the front, beyond the plane x_2, the corresponding quantities are u_2, ρ_2, and T_2. Between x_1 and x_2, ρ and u are functions of x, and T may not be relevant, since the gas may be far from thermal equilibrium in the shock front.

To determine the three ratios u_2/u_1, ρ_2/ρ_1, and T_2/T_1 we need three relationships, or "jump conditions," across the front. Two of these are provided by the conservation of matter and momentum, obtained by integrating equations (5-2) and (5-1) across the front

$$\rho_1 u_1 = \rho_2 u_2 \tag{5-15}$$

$$p_1 + \rho_1 u_1^2 = p_2 + \rho_2 u_2^2 \tag{5-16}$$

In deriving equation (5-16) we have set **B** and $\nabla \phi$ equal to zero in equation (5-1), and made use of equation (5-15). The temperature of the gas is determined by the conservation of energy, including

5. DYNAMICS OF THE INTERSTELLAR GAS

the effects of radiation as well as compression. We consider here the two limiting cases in which the gas does not radiate at all and in which radiation is so rapid that radiative equilibrium is established.

In the first case the gas is compressed adiabatically, and p varies as ρ^γ. To obtain the third jump condition we multiply equation (5-1) by u, defined as the velocity v_x in a reference frame moving with the shock; integrating across the front, we find

$$\frac{\gamma}{\gamma-1}\frac{p_1}{\rho_1}+\frac{1}{2}u_1^2=\frac{\gamma}{\gamma-1}\frac{p_2}{\rho_2}+\frac{1}{2}u_2^2 \qquad (5\text{-}17)$$

If equations (5-15), (5-16), and (5-17) are combined we obtain [1], after some algebra

$$p_2=\frac{2\rho_1 u_1^2}{\gamma+1}+\frac{\gamma-1}{\gamma+1}p_1 \qquad (5\text{-}18)$$

$$\frac{\rho_1}{\rho_2}=\frac{\gamma-1}{\gamma+1}+\frac{2}{\gamma+1}\frac{\gamma p_1}{\rho_1 u_1^2} \qquad (5\text{-}19)$$

For a strong shock, u_1 much exceeds $(\gamma p_1/\rho_1)^{1/2}$, the adiabatic sound velocity in front of the shock, and the last terms on the right-hand sides of these two equations can be ignored. In this case ρ_2/ρ_1 is equal to its upper limit, which for a monotonic gas ($\gamma = 5/3$) equals 4; however, both p_2 and T_2 increase linearly with u_1^2, the square of the shock front velocity relative to the gas in front.

We consider next the second case, in which radiative equilibrium is established. Under typical conditions prevailing in an interstellar H I cloud the cooling time given in Table 4.11 is about 10^4 years for a density of some 20 H atoms/cm^3, corresponding to a strong shock passing through a standard cloud. Hence T will reach its equilibrium value within approximately 0.1 parsec behind the shock front. One may regard x_2 as the value of x at which this equilibrium has been reached, and include in the shock front between x_1 and x_2 not only the abrupt initial increase of temperature and density discussed above but also the subsequent cooling and compression, as indicated in Figure 5.1. If the hydrogen gas is neutral on both sides of the front, T will be essentially the same

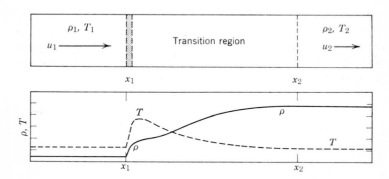

Fig. 5.1. Schematic diagram of radiating shock. In the upper diagram the fluid is shown moving from left to right. At x_1 the fluid enters the shaded area, representing the nonradiative shock, followed by the transition region, where the temperature drops by radiation. The changes of density and temperature are indicated schematically in the lower figure.

both at x_1 and at x_2, and in this sense the shock may be regarded as isothermal. In this case p equals ρC^2, and combination of equations (5-15) and (5-16) yields

$$\frac{\rho_2}{\rho_1} = \frac{u_1^2}{C^2} \qquad (5\text{-}20)$$

where the sound speed, C, is given in equation (5-6). If u_1 is 10 km/sec, an appropriate value for u_1 when two clouds collide, and if C is about 1 km/sec, a reasonable value for an H I cloud, ρ_2 is about 100 ρ_1. Evidently large compressions are possible in an isothermal shock.

We consider next the modifications resulting when a magnetic field is present; as we shall see, the compression in an isothermal shock can be much reduced in this case. If we assume that the lines of force are straight and are everywhere perpendicular to the x axis, the **B** · **∇B** term disappears from equation (5-1), and inte-

5. DYNAMICS OF THE INTERSTELLAR GAS

grating this equation across the front yields, in place of equation (5-16),

$$p_1 + \rho_1 u_1^2 + \frac{B_1^2}{8\pi} = p_2 + \rho_2 u_2^2 + \frac{B_2^2}{8\pi} \qquad (5\text{-}21)$$

We have already seen that the magnetic flux remains constant through any circuit moving with a conducting fluid, and for the one-dimensional compression across a shock we have, therefore,

$$\frac{B_1}{\rho_1} = \frac{B_2}{\rho_2} \qquad (5\text{-}22)$$

In a strong nonradiating shock we have seen that if **B** is ignored, the density increases by only a factor 4, and the magnetic pressure, $B^2/8\pi$ increases by only a factor 16; if B is less than 3×10^{-6} G in front of the shock, the magnetic pressure behind the shock is less than one-fifth the value of p_2 found from equation (5-18), if ρ_1 and u_1 have typical values for interstellar clouds. The magnetic field in this case has no very great effect on the shock front, and the neglect of **B** is valid.

For an isothermal shock p equals ρC^2, with the same value of C on both sides of the front, and equations (5-21), (5-22), and (5-15) can be combined to give ρ_2/ρ_1. In the limiting case where the Alfvén velocity, V_A, much exceeds C, and ρ_2 much exceeds ρ_1, we obtain

$$\frac{\rho_2}{\rho_1} = 2\tfrac{1}{2} \frac{u_1}{V_A} \qquad (5\text{-}23)$$

The criterion for a strong supersonic shock is now that u_1 must be large compared to V_A rather than compared to C. If B is 3×10^{-6} G and ρ_1 in an H I cloud is 2.4×10^{-23} g/cm^{-3} (see Table 5.1), V_A computed from equation (5-10) is 1.7 km/sec, somewhat exceeding C; the compression given by equation (5-23) is then substantially less than is found from equation (5-20).

When energetic particles are present, gyrating around the lines of force, the pressure, p_R, contributed by these suprathermal particles must be included in equation (5-21). For a given u_1 the value of ρ_2/ρ_1 may be somewhat reduced by this additional effect.

Parameters of the Interstellar Medium

In the subsequent discussions of interstellar dynamics, specific values for the parameters of the medium must be assumed in the calculations. Table 5.1 gives values which have been adopted for

TABLE 5.1

Parameters of the Interstellar Gas

Mean density, ρ	2×10^{-24} g/cm^3
Particle density of hydrogen in clouds	10/cm^3
Root mean square cloud velocity $\langle v^2 \rangle^{1/2}$	14 km/sec
Isothermal sound speed, C	
H I region at 50°K	0.57 km/sec
H II regions at $10^{4°}$K	11 km/sec
Magnetic field, B	3×10^{-6} G

the more important quantities on the basis of the observational data discussed earlier.

The mean density of 2×10^{-24} g/cm^3 corresponds to an average value of n_H of 0.85 per cm^3, as compared to the average value of 0.5 cm^{-3} obtained directly from the 21-cm data for the galactic disc at distances between 5 and 10 kpc from the center, together with a helium/hydrogen ratio of 0.1 by particle number (see Table 4.7) corresponding to a fractional abundance, Y, of helium, by mass, equal to 0.29. This 70% increase allows for some additional density within spiral arms. Further corrections for saturation of the 21-cm emission line and possibly also for the presence of molecular H_2 might increase this average value substantially, though probably not by a factor of 5 or more (see Section 1.2). The value of n_H within clouds is equal to that adopted for the standard cloud in Table 3.4. The same value has been taken as a typical initial density within an H II region. While this value is probably representative, densities differing by orders of magnitude are known to be present in interstellar space.

5. DYNAMICS OF THE INTERSTELLAR GAS

The r.m.s. cloud velocity is obtained from velocity dispersions measured both in the 21-cm emission line and in the visible absorption lines of Ca II. The r.m.s. radial velocity indicated in Chapter 2 is about 8 km/sec, and 14 km/sec is obtained with the arbitrary assumption that the velocity distribution is isotropic. The observed differences between Na I and Ca II lines and between the emission and absorption lines at 21 cm are not inconsistent with the view that the more massive, colder clouds have lower velocities, and that a weighting of cloud velocities by mass would give a lower value of $\langle v^2 \rangle^{1/2}$ than the one adopted. On the other hand, the amount of interstellar gas moving with radial velocities much greater than 25 km/sec is enormously greater than would be expected for a Maxwellian distribution of velocities with the adopted r.m.s. value.

The temperature of 50° adopted for H I regions is less than the values normally taken, which range from 100 to 125°. The discussion in Section 4.4 indicates that there are serious theoretical difficulties in understanding how H I clouds can be maintained at a temperature as great as 100°; the observational evidence in Sections 2.3 and 2.4 is entirely consistent with a lower temperature for at least the denser, more massive clouds. For H II regions the evidence suggests a temperature more nearly 7000° than 10,000°K, but this small change is not as yet conclusively indicated. So many observational results for H II regions have been interpreted on the basis of a 10,000° temperature that adoption of a different temperature at this time would not seem warranted. The values of the sound velocity, C, in Table 5.1 are computed with μ/m_H equal to 1.27 and 0.67 in H I and H II regions, respectively, with helium atoms taken to be neutral.

The value of 3×10^{-6} G adopted for B is less than the upper limit of 5×10^{-6} G obtained from attempts to measure the Zeeman effect in 21-cm absorption lines, and is not inconsistent with the measurements of Faraday rotation (see Sections 2.3 and 2.4). Theoretical arguments presented below indicate that a greater field would be very difficult to reconcile with star formation from the interstellar gas, while a much smaller field would increase the difficulties associated with the hypothesis that cosmic rays are chiefly confined to the Galaxy. Adoption of this value implies that the galactic radio emission must originate almost entirely in regions where the relativistic particle density much exceeds the value observed at the Earth, or where the magnetic field much exceeds its mean value, or both.

Energy Balance in Cloud Collisions

It seems relatively well-established that the presence of supersonic velocities in the interstellar gas must imply a dissipation of energy. Here we attempt to evaluate the power loss and to survey the possible energy sources available. The detailed mechanisms involved in accelerating the gas are discussed in subsequent sections.

To compute the dissipation one may adopt the simple model in which separate clouds are moving at random and collide at intervals. We denote by P_c the power dissipated per cm^3 in such cloud collisions. Since the r.m.s. relative velocity is $2^{1/2} \times 14$ or 20 km/sec, the observed number of 8 clouds per kpc gives a collision between an element of one cloud and some other cloud in 6.3×10^6 years. Since this relative velocity much exceeds either the sound speed, C, of 0.5 km/sec or the Alfvén velocity of 1.7 km/sec through the undisturbed cloud material, strong shocks will be produced at the collision, and virtually all the kinetic energy of relative motion will be radiated away. The values given in Table 5.1 give a density of kinetic energy equal to 2×10^{-12} ergs/cm^3. On the average the kinetic energy in relative motion is half the total kinetic energy of two colliding clouds, and with one collision every 6.3×10^6 years we obtain

$$P_c = 5 \times 10^{-27} \text{ ergs cm}^{-3} \text{ sec}^{-1} \qquad (5\text{-}24)$$

The uncertainties in the basic parameters adopted indicate that P_c may well differ by half an order of magnitude from the value in equation (5-24). The power taken away from the kinetic energy of the clouds is presumably converted into infrared radiation from

5. DYNAMICS OF THE INTERSTELLAR GAS

the ions C^+, Si^+, and Fe^+, and possibly also from H_2, and in principle could be measured directly.

To offset this power dissipation requires a substantial supply of energy. The two most likely energy sources are the kinetic energy of photoelectrons produced by ultraviolet radiation from early-type stars and the kinetic energy of supernova shells. Values of P_u and P_s, the average power per cm³ released in the interstellar medium from these two sources, are estimated below.

If we assume that all stellar photons shortward of the Lyman limit are absorbed in H II regions, and do not escape from the Galaxy, the thermal power input, P_u, is defined as the total number of such ultraviolet photons emitted by all the early-type stars, multiplied by the energy per photoelectron given in equation (4-78), and divided by the volume occupied by the interstellar gas. The physical processes by which this heat energy in H II regions is converted in part into kinetic energy of H I clouds are discussed in Section 5.3. Here we compute the heat energy available.

Instead of determining P_u by adding up the contributions from stars of each spectral type, which involves uncertainties both in the intrinsic ultraviolet luminosity and in the stellar luminosity function, we shall use instead the measured values of the continuous thermal radio emission at high frequencies. The mean power radiated at these frequencies per cm³ is directly proportional to n_e^2 and hence to the rate of electron-ion recombination per cm³, with a constant of proportionality which depends on the temperature [see equations (2-46) and (4-57)]. Since the recombination rate, in turn, must equal the corresponding number of ultraviolet photons absorbed per cm³ per sec, and since the mean heating energy per absorbed stellar photon can be estimated, P_u can be computed directly.

Specifically, we may take the result, given in Section 2.2, that the observed intensity of radiation at 21.6 cm corresponds to a value of 0.20 cm⁻⁶ for $\langle n_e^2 \rangle$, the mean square electron density in the galactic plane at distances between 7 and 10 kpc from the center. According to Section 4.3 the effective number of recombinations per second per cm³, which equals the corresponding number of photoionizations produced by starlight, is equal to $\langle n_e^2 \rangle$ multiplied by the partial recombination coefficient, $\alpha^{(2)}$. According to Section 4.4 the average energy per emitted photoelectron is given by equation (4-78), on the assumption that essentially all the ultraviolet radiation from the star is absorbed. Hence P_u is given by

$$P_u = \langle n_e^2 \rangle k T_c \psi_3(\beta_c) \sum_2^\infty \alpha_j$$
$$= 5.7 \times 10^{-28} \frac{T_c \psi_3(\beta_c) \phi_2(\beta)}{T^{1/2}} \quad (5\text{-}25)$$

For an electron temperature of $10^{4}\,°\mathrm{K}$, $\phi_2(\beta)$ equals 1.25. The value of $T_c \psi_3(\beta_c)$ should be averaged over the ultraviolet luminosities of the early-type stars. We shall here set T_c equal to 35,000°, which seems reasonably representative of an estimated temperature range from 20,000° to 55,000° for spectral types between B0 and O5; from Table 4.8 we then find that $\psi_3(\beta_c)$ equals 1.42. Equation (5-25) then yields

$$P_u = 3.5 \times 10^{-25} \text{ ergs cm}^{-3} \text{ sec}^{-1} \quad (5\text{-}26)$$

Since P_u is about 2 orders of magnitude greater than P_c in equation (5-24), a conversion efficiency of about 1% in converting thermal energy of H II regions into expansion velocities will suffice to maintain the interstellar velocities.

The value of P_s, the rate of kinetic energy released by supernovae, is equal to the kinetic energy per explosion, multiplied by the number of explosions per galaxy per second, divided by the volume of the galactic disc; in this very rough calculation we assume that all supernova shells dissipate their energies in the galactic disc. For a type I supernova, with an initial expansion velocity of some 20,000 km/sec, the initial energy of the ejected shell has been estimated as roughly about 10^{51} ergs [4], some 30 times the total energy radiated as visible light. For supernovae of type II the expansion velocity is only about 6000 km/sec, and since the ejected mass is completely uncertain we shall neglect these objects.

5. DYNAMICS OF THE INTERSTELLAR GAS

The frequency of supernovae is highly uncertain, but one explosion of each type every 100 years within the Galaxy would not be inconsistent with present information [5].

If the galactic disc is taken to be a cylinder with a radius of 10,000 pc and a thickness of 200 pc, we obtain

$$P_s = 1.7 \times 10^{-25} \text{ ergs cm}^{-3} \text{ sec}^{-1} \tag{5-27}$$

If the energy released in a supernova of type II equals that in type I, this estimate should be doubled; on the other hand, if it is considered that type I supernovae are mostly outside the galactic disc, and that only half the ejected matter strikes the disc, the value in equation (5-27) should be halved. Estimates of the supernova energy and frequency introduce even larger uncertainties into equation (5-27).

One cannot exclude the possibility that other souces of energy may also contribute to the maintenance of the cloud velocities. For example, the rotational energy of the Galaxy must be considered. If the mean density given in Table 5.1 is again adopted, galactic rotation of the gas at 250 km/sec provides a kinetic energy of about 6×10^{-10} ergs/cm^3, sufficient to maintain P_c for only 4×10^9 years. If some mechanism could be found for using the rotational energy of the Galaxy as a whole, including the stars, for maintaining the random velocities of the clouds, this energy reservoir could be of great importance. Energy output in unobserved forms from supernovae and from other objects may conceivably be significant.

5.2 Equilibrium Density Distribution

Before discussing the detailed dynamical processes occurring in the interstellar medium, we consider the density distribution as determined by the velocities of the clouds, by the magnetic field, and by the relativistic particles embedded in this field. The discussion is not conclusive, but does restrict somewhat the number of possible theoretical models for the interstellar medium.

Galactic Distribution

The galactic density distribution of the gas, together with the magnetic field and the relativistic particles coupled to it, may be discussed on the basis of the virial theorem, equation (5-11). In this relation d^2I/dt^2 may here be set equal to zero, on the assumption of a nearly stationary state, and the integrals may be extended over all the gas within a galaxy.

The simple case of a spherical galaxy, with no systematic motions, no magnetic field, and no relativistic particles will be considered first. In this case, $3U$ must equal $-W$; if we assume that the gravitational force is due to a spherical distribution of stars with a uniform density ρ_S, we obtain

$$r_G^2 = \frac{3\langle w_G^2 \rangle}{4\pi G \rho_S} \tag{5-28}$$

where r_G denotes the radius of the gas, defined as the r.m.s. value of r, averaged over the total mass of gas, and $\langle w_G^2 \rangle$ denotes the mean square random velocity of particles in the gas. If r_G is much less than r_S, the corresponding r.m.s. radius of the stellar system, the use of a constant ρ_S in the computation of W is a good approximation. For the stars a relation similar to equation (5-28) may also be written, with a numerical constant that depends on the stellar density distribution. Dividing these results we have the approximate relation

$$\frac{r_G^2}{r_S^2} \approx \frac{\langle w_G^2 \rangle}{\langle w_S^2 \rangle} \tag{5-29}$$

For a gas at densities of a few atoms per cm^3 or more, $\langle w_G^2 \rangle^{1/2}$ is not likely to much exceed 10 km sec^{-1}. Stellar velocities in galaxies are generally 1 order of magnitude greater, and it follows that a system of gas in hydrostatic equilibrium within a spherical galaxy, or in the spherical nucleus of a galaxy, must be concentrated towards the center, unless its temperature is somehow much higher than normal, or unless cosmic rays and magnetic fields have a total energy much exceeding U_G. If ρ_G and ρ_S are comparable,

5. DYNAMICS OF THE INTERSTELLAR GAS

the self-gravitational energy of gas must be taken into account in computing W, and r_G is decreased. In this case it is readily shown [6] that if $\langle w_G^2 \rangle$ is fixed there is a maximum mass of gas that can be in equilibrium at the center of such a system. In globular clusters $\langle w_G^2 \rangle$ and $\langle w_S^2 \rangle$ can be about equal, in which case the gas escapes by evaporation.

Rotation of a galaxy can change these conclusions substantially, since centrifugal force can support the gas against gravity in the plane of rotation. In this case the kinetic energy term, $2T$, may cancel most of the gravitational term, W, in the virial theorem. To take this cancellation into account we compute $2T + W$ when v_θ is expressed in the form

$$v_\theta = \left(r \frac{\partial \phi}{\partial r} \right)^{1/2} - v_L \qquad (5\text{-}30)$$

where we have introduced the cylindrical coordinates r, θ, and z; the gravitational potential, ϕ, is assumed to be a function of r and z only. The first term on the right-hand side is the "circular velocity" at which the centrifugal force just cancels the gravitational force towards the axis of rotation; the velocity v_L is the lag in rotation behind the circular velocity. If equation (5-30) is substituted into the defining equation for T in set (5-12), we obtain [3, 7]

$$2T + W = \int \rho \, dV \left\{ v_L^2 - 2v_L \left(r \frac{\partial \phi}{\partial r} \right)^{1/2} - z \frac{\partial \phi}{\partial z} + v_r^2 + v_z^2 \right\} \quad (5\text{-}31)$$

Evidently $2T + W$ can be zero or even negative; in contrast to the spherical case the presence of rotation makes it possible for relatively cold gas to be in equilibrium even at large distances from the galactic center.

In applying these results to the Galaxy we cannot ignore the pressure of cosmic rays, which we denote by p_R. For relativistic particles the pressure is one-third the energy density, and from equation (2-66) we have

$$p_R = 0.4 \times 10^{-12} \text{ dynes cm}^{-2} \qquad (5\text{-}32)$$

The thermal pressure of the gas is about 1 order of magnitude less, but the turbulent pressure, associated with cloud velocities, is somewhat greater in the galactic midplane (see below). The magnetic energy density $B^2/8\pi$ is uncertain, but if B is 3×10^{-6} G, this energy density is about equal to the value of p_R above. There is no information on how p_R and $B^2/8\pi$ vary with distance from the Sun, and their contribution to the virial theorem depends entirely on this variation. It is possible that $p_R + B^2/8\pi$ may be constant throughout the Universe. The surface terms in p_R and B will then cancel out the volume terms, leaving the density distribution of gas unaffected by cosmic rays and magnetic forces. This assumption has the consequence that the total energy in the Universe associated with cosmic rays and the magnetic field is very large and must be assumed primordial.

An alternative possibility is that both p_G and $B^2/8\pi$ are much smaller outside our Galaxy than inside, and that the surface terms may be ignored in equation (5-11) if an integral is taken over the entire system. We analyze the consequences of such an assumption, considering first that cosmic rays and magnetic field are confined to a halo region, extending some 10 kpc from the galactic center, and next that confinement occurs in the galactic disc.

If cosmic rays are assumed to fill a region as large as the galactic halo, with a radius of some 10,000 pc, the term U in equation (5-11) becomes large, and must be offset by a substantial lag velocity for the interstellar gas. For example, if p_R is constant out to about 10 kpc from the galactic plane, the value of v_L required [7] in equation (5-31) is about 10 to 15 km sec^{-1}, depending on parameters chosen for galactic rotation and for the amount of interstellar gas. Such a large lag velocity seems inconsistent with the radial velocity measures of the interstellar Ca II lines (see Section 2.3). This lag velocity may be reduced if sufficient mass is assumed in the halo to provide a large negative gravitational energy that is not offset by rotation; the required density is about 10^{-2} H atoms per cm^3, corresponding to a mass of some 10^9 suns. While there is no direct evidence for so much material surrounding the Galaxy, this possibility cannot be excluded.

5. DYNAMICS OF THE INTERSTELLAR GAS

Distribution Perpendicular to the Galactic Plane

The situation where cosmic rays and magnetic fields are mostly confined to the galactic disc is analyzed here in some detail, since the theoretical predictions may be compared with the observations on the density distribution for the various components of the interstellar gas. As pointed out in Sections 2.2, 2.3, and 3.2, neutral H atoms, Ca II ions, and dust grains are all confined to a galactic layer some 200 pc thick. We investigate the density distribution in this layer not with the virial theorem but rather with the equation of hydrostatic equilibrium, obtained from equation (5-1) if the time derivatives on the left-hand side are set equal to zero.

As we have seen above, the total material pressure $p(z)$ is the sum of the cosmic ray pressure, $p_R(z)$, and the gas pressure, $p_G(z)$; both pressures are taken to be functions of z, the distance from the galactic plane. The value of $p_R(0)$ is given in equation (5-32). Since the random cloud velocities much exceed the random thermal velocities in H I regions, the thermal pressure of the gas may be neglected compared to the macroscopic turbulent pressure associated with cloud motions, and for z equal to 0, p_G becomes

$$p_G(0) = \tfrac{1}{3}\rho\langle v^2\rangle = 1.3 \times 10^{-12} \text{ dynes cm}^{-2} \quad (5\text{-}33)$$

where the mean gas density, ρ, and the mean square cloud velocity $\langle v^2 \rangle$ have been evaluated from Table 5.1. From equations (5-32) and (5-33) it follows that the ratio of p_R to p_G, which we denote by α, equals 0.3.

The two magnetic terms in equation (5-1) depend on the topography as well as the strength, **B**, of the magnetic field. If the lines of force are assumed straight and parallel to the galactic plane, then the $\mathbf{B} \cdot \nabla \mathbf{B}$ term, which represents the net force produced by the tension along **B**, vanishes, and the magnetic field contributes an additional pressure, which may be written as

$$p_B(z) = \frac{B^2}{8\pi} \quad (5\text{-}34)$$

The value of B is uncertain, but if we assume arbitrarily that β, the ratio of p_B to p_G, is equal to α, then B equals 3×10^{-6} G.

Since we are considering a situation where p_G, p_R, and p_B all go to zero together with increasing z, we may assume that α and β are constants. Equation (5-1) then becomes [8]

$$(1 + \alpha + \beta)\frac{1}{3\rho}\frac{d}{dz}(\rho\langle v^2\rangle) = -\frac{d\phi}{dz} = g_z \qquad (5\text{-}35)$$

We assume that g_z varies linearly with z with a slope dg_z/dz, in rough agreement with the observed data in Figure 1.1. If $\langle v^2\rangle$ is assumed constant, equation (5-35) has the solution

$$\rho(z) = \rho(0)e^{-(z/h)^2} \qquad (5\text{-}36)$$

where

$$h = \left\{\frac{2(1 + \alpha + \beta)\langle v^2\rangle}{-3 dg_z/dz}\right\}^{1/2} = 130(1 + \alpha + \beta)^{1/2} \text{ pc} \qquad (5\text{-}37)$$

The numerical value in equation (5-37) has been determined with dg_z/dz equal to -2.5×10^{-11} cm sec^{-2} pc^{-1}. The effective thickness, $2H$, of the interstellar layer is defined as the total mass per cm^3 divided by $\rho(0)$, giving

$$2H = \pi^{1/2}h = 230(1 + \alpha + \beta)^{1/2} \text{ pc} \qquad (5\text{-}38)$$

If α and β were both zero, or if p_R and p_B were constant with z through the galactic disc, $2H$ would equal 230 pc, in reasonable agreement with the values of 220, 240, and 190 pc found for H atoms, Ca$^+$ ions, and grains, respectively. If α and β are each set equal to 0.3 in equation (5-38), $2H$ equals 290 pc, a significantly greater value, but one which is probably not excluded by the observations. If this value of $2H$ seems somewhat large, the data could also be made consistent with the assumed value of $\alpha + \beta$ by decreasing $\langle v^2\rangle^{1/2}$ to about 10 km/sec, which would be consistent with observations. However, to avoid decreasing p_G and thus increasing α, as $\langle v^2\rangle$ is decreased, a corresponding increase in $\rho(0)$, the gas density in the galactic midplane, by a factor about 2 must

5. DYNAMICS OF THE INTERSTELLAR GAS

be assumed. This modification also cannot be excluded at the present time.

We may conclude that a restriction of magnetic and cosmic ray energies to the galactic disc seems consistent with the data, provided β does not much exceed the assumed value of 0.3. If B equalled 2×10^{-5} G, the value suggested by the observed synchrotron radiation, β would be increased to about 12, and the confinement of so great a magnetic field to the galactic disc would seem excluded. A low field confined in the galactic disc may be unstable, with loops of force rising out of the galactic plane under the outwards pressure of the relativistic particles [8]. However, the exact conditions under which this instability would occur are uncertain, as are the full effects which such an instability would produce.

Another uncertainty in this analysis of pressure equilibrium is the effect produced by streams of hydrogen gas falling into the galactic plane. As noted in Section 2.2 the observed mean value of $n(\text{H I})L$ for the rapidly approaching hydrogen clouds at high latitude is about 10^{19} cm^{-2} with a mean radial velocity of 75 km/sec. If L is assumed to equal 1 kpc, the mean pressure $n_\text{H} m_\text{H} v^2$ associated with the retardation of this stream becomes 3.1×10^{-13} dynes/cm^2, about one-fourth p_G. If the effective L were less than 1 kpc this contribution to the momentum balance in the z direction would be correspondingly increased, and the conclusions reached in the paragraphs above would require substantial changes.

Distribution in the Galactic Disc

We consider now the variation of gas density within the galactic disc. As pointed out in Section 3.4, the grains are concentrated in individual clouds, and the gas also tends to be concentrated in such clouds. We discuss here several points of view that have been put forward to explain this observed density concentration.

According to one explanation the clouds are in pressure equilibrium with the gas between the clouds. If such equilibrium is absent, a cloud will tend to expand at the sound velocity, amounting to about 0.6 km/sec (see Table 5.1). If the cloud radius is initially

7 pc, its volume will increase by a factor 10 in about some 10^7 years. The continued existence of the cloud is readily understood if the rarefied medium between the clouds is at a sufficiently high temperature to establish pressure equilibrium. If, for example, the cloud temperature were 50°, with the intercloud medium at 2000°, a ratio of densities of 40 to 1 between the cloud and the intercloud medium would be required for equal pressures; the cloud density in Table 5.1 would then correspond to a value of about 0.3 cm^{-3} for n_H between the clouds. The computed temperatures in Table 4.10 are not consistent with this hypothesis; even for the maximum cosmic-ray ionization rate the temperature increases less rapidly than $1/n_H$ as the density decreases. However, there are a number of uncertainties in the theoretical computations of the temperature in H I regions, especially at low densities. In particular a lower ratio of heavy atoms to H atoms and increased efficiency of cosmic ray heating at low densities might materially increase T at low n_H. Other energy sources, such as scattered stellar ultraviolet light and hydrogen clouds streaming at high velocity towards the galactic plane, might also increase the temperature of the low density regions. Evidently the possibility of pressure equilibrium between the clouds and the intercloud medium cannot be excluded.

According to another explanation [9] the clouds are not in pressure equilibrium, but are compressed in occasional cloud–cloud collisions, with free expansion between collisions. On this picture the mean pressure on a single cloud equals p_G, the pressure due to the random motion of all the other clouds. If we equate $n_c kT$, the mean thermal pressure within the cloud, to p_G in equation (5-33), we obtain

$$n_c = \frac{\rho \langle v^2 \rangle}{3kT} = 200 \text{ cm}^{-3}, \qquad (5\text{-}39)$$

a rather high value for n_c, the total particle density within the cloud. It is not obvious under what conditions fluid motions at supersonic velocities will produce the large-scale density inhomogeneities observed in the interstellar medium.

5. DYNAMICS OF THE INTERSTELLAR GAS

It is entirely possible that magnetic forces may play an important role in shaping the interstellar clouds [8]. Uncertainty as to magnetic topography makes it difficult to decide between the many different possibilities in this connection.

5.3 Expansion of H II Regions

A primary characteristic of H II regions is their transitory character. The lifetime of an O-type star is at most 10^7 years, about equal to the interval between collisions for an interstellar cloud. During this time the interstellar gas will travel a distance comparable to the diameter of an H II region, since at 10 km/sec the distance traveled in 10^7 years is 100 pc.

When an O star is formed, it is presumably surrounded by relatively dense interstellar gas. Before star formation began this gas may be assumed neutral and cold. The formation of the O star will produce two effects. First, the gas surrounding the new star will become ionized. Since the mean free path of an ultraviolet photon is very short in neutral hydrogen, the photons will travel freely through the ionized gas immediately surrounding the star, but will be absorbed in a relatively thin surrounding shell of neutral hydrogen, producing new ionization. Thus the ionized and neutral gases are separated by an ionization front, which moves rapidly outward as more and more atoms become ionized by the stream of photons. Second, the heated gas will be at a much higher pressure than the surrounding cool gas, and will tend to expand. Since the expansion velocity is likely to exceed the sound velocity in the surrounding H I region, a shock wave may be expected to form, moving out through the neutral gas. The dynamical analysis of H II regions must consider the interactions between the ionization front and the shock front, together with the equations of motion for the gas behind the two fronts.

The relevant equations for an ionization front, analogous to equations (5-15) through (5-20) for a shock front, are derived below. These relations are then used in a discussion of the dynamical evolution of an H II region.

Ionization Fronts

As in the discussion of shock fronts, we consider a one-dimensional front propagating through a homogeneous medium with constant velocity, u_1, and we adopt a frame of reference moving at this same velocity. Figure 5.1 is again applicable, except that now between x_1 and x_2 there is not only a change in velocity and density but also a change of ionization state and kinetic temperature produced by absorption of ultraviolet photons. The thickness of the ionization front must somewhat exceed the mean free photon path, given in equation (4-49); if the change in density of neutral H within the front is ignored, τ_v equals 1 for a path length of $0.05/n(\text{H I})$. The distance traveled by the front during the time, t_T, required to establish thermal equilibrium in the H II region behind the front may be computed from equation (4-80), and determines the thickness of the transition region; this distance equals about $0.2/n_p$ pc if the front velocity is 10 km/sec and if n_p is taken to be constant.

Equations (5-15) and (5-16) are valid as before, with the one difference that the mass flow through the front is no longer arbitrary. Instead, the number of H atoms flowing through the front per unit area per second must equal J, the corresponding number of ionizing photons reaching the front. Hence equation (5-15) becomes

$$\rho_1 u_1 = \rho_2 u_2 = \mu_i J \qquad (5\text{-}40)$$

where μ_i is defined as the mean mass of the gas per newly created positive ion; if the abundance of helium relative to hydrogen is again taken to be 0.10 by particle number (see Table 4.7), μ_i equals $1.40\ m_\text{H}$ if the helium remains neutral, and $1.27\ m_\text{H}$ if the helium becomes singly ionized in the front. In terms of $S_u(r)$, the net outwards flow of ultraviolet photons per second through a shell of radius r, J is given by

$$J = \frac{1}{4\pi r^2} S_u(r) \qquad (5\text{-}41)$$

In an isothermal shock front the sound speeds C_1 and C_2 ahead of and behind the front are equal. In an ionization front C_2 and

5. DYNAMICS OF THE INTERSTELLAR GAS

C_1 are markedly different, each value being determined by the equilibrium temperature and mean molecular weight in H II and H I regions (see Table 5.1). The general solution of equations (5-15) and (5-16) together with (5-6) is

$$\frac{\rho_2}{\rho_1} = \frac{(C_1^2 + u_1^2) \pm \{(C_1^2 + u_1^2)^2 - 4u_1^2 C_2^2\}^{1/2}}{2C_2^2} \quad (5\text{-}42)$$

The condition that equation (5-42) gives a real value places a restriction on u_1. Either u_1 must exceed an upper critical value, u_R, given by

$$u_R = C_2 + (C_2^2 - C_1^2)^{1/2} \approx 2C_2 \quad (5\text{-}43)$$

or u_1 must be less than a lower critical value, u_D, given by

$$u_D = C_2 - (C_2^2 - C_1^2)^{1/2} \approx \frac{C_1^2}{2C_2} \quad (5\text{-}44)$$

The approximate equality in equations (5-43) and (5-44) holds because C_2 exceeds C_1 by about 1 order of magnitude in an interstellar ionization front.

The nature of the ionization front depends on the value of u_1, which is determined by equation (5-40). If u_1 exceeds u_R, the solution is called "R type"; R refers to a rarefied gas, since for ρ_1 sufficiently low, u_1 will always exceed u_R. In this case ρ_2 exceeds ρ_1, and u_1 exceeds $2C_2$. If we take the negative sign in equation (5-42), the motion of the front is supersonic relative to the gas behind as well as ahead of the front. This type of front is referred to as "weak R", since the relative change of density is small. The "strong R" fronts, corresponding to the positive sign in equation (5-42), and to a large density increase across the front, are not usually of interest for interstellar conditions; there is no mechanism for maintaining the large outwards velocity of the compressed hot gas that would be required if the ionization front were of strong R type. For a weak R front, in which u_1 much exceeds C_2, the changes in density and relative velocity across the front are given by

$$\frac{\rho_2}{\rho_1} = 1 + \frac{C_2^2}{u_1^2} \tag{5-45}$$

$$u_1 - u_2 = \frac{C_2^2}{u_1} \tag{5-46}$$

to first order in C_2^2/u_1^2.

If u_1 is between u_R and u_D, an ionization front of this simple type is not possible. In this case a shock wave must precede the ionization front, reducing u_1 to u_D or below. If u_1 is exactly equal to u_D, the front is called "D-critical," u_2 is exactly equal to C_2 and the density change is given by

$$\frac{\rho_2}{\rho_1} = \frac{C_1^2}{2C_2^2} \tag{5-47}$$

giving p_2 equal to $p_1/2$. More generally, under interstellar conditions the shock ahead of the ionization front will compress the gas sufficiently so that u_1 found from equation (5-40) is less than u_D; this solution is called "D type", with D referring to a dense gas. In a D-type front ρ_2 is always less than ρ_1. The "weak D" front, with the smaller relative decrease in density, now corresponds to the positive sign in equation (5-42). For these fronts, which are subsonic with respect to the gas on both sides, the density and velocity changes are given by

$$\frac{\rho_2}{\rho_1} = \frac{u_1}{u_2} = \frac{C_1^2}{2C_2^2} \xi \tag{5-48}$$

where ξ, given by

$$\xi \equiv 1 + \left(1 - \frac{4u_1^2 C_2^2}{C_1^4}\right)^{1/2} = 1 + \left(1 - \frac{u_2^2 \xi^2}{C_2^2}\right)^{1/2} \tag{5-49}$$

varies from 1 to 2 as u_2 decreases from C_2 to zero. In these equations we have neglected terms of order u_1/C_1; in accordance with equation (5-44) and the condition that u_1 is less than u_D, such terms are less than $C_1/2C_2$.

5. DYNAMICS OF THE INTERSTELLAR GAS

Strong D fronts, in which u_2 exceeds C_2, may also occur, but their existence is limited to a relatively brief interval in the development of an H II region, and they will not be considered here.

Initial Ionization of the Gas

When an O star is first formed, the ultraviolet luminosity may be expected to rise rather abruptly. For example, calculations for a model star of $30M_\odot$, as it approaches the main sequence, show that during the final increase of $S_u(0)$, the emission rate of photons beyond the Lyman limit, this quantity doubles in 4000 years [10]. Thus the initial growth of an H II region may be computed on the assumption that the central star abruptly starts to shine at its full luminosity.

Close to the star the ionization front will move away at a very rapid speed. Consider, for example, an ionization front at a distance of 5 pc from an O7 star, with a hydrogen density of 10 atoms per cm^3. From equations (5-40) and (5-41), with the numerical value of $S_u(0)$ for an O7 star in Table 4.5, we find that u_1 equals 910 km/sec. Evidently u_1 exceeds u_R, and the front is of weak R type. If r_i denotes the radius of the ionization front we may write

$$\frac{dr_i}{dt} = \frac{S_u(r_i)}{4\pi n_H r_i^2} = \frac{1}{4\pi n_H r_i^2}\left\{ S_u(0) - \frac{4}{3}\pi r_i^3 n_H^2 \alpha^{(2)} \right\} \quad (5\text{-}50)$$

where the effect of absorption on S_u has been evaluated by means of equation (4-54), and where the helium atoms have been assumed neutral. An integration yields

$$r_i^3 = r_S^3\{1 - \exp(-n_H \alpha^{(2)} t)\} \quad (5\text{-}51)$$

where equation (4-56) has been used for r_S, the equilibrium radius of the H II region at the density n_H, which we have assumed uniform throughout the region.

Equations (5-50) and (5-51) are valid only as long as dr_i/dt exceeds u_R or about $2C_2$ (see equation [5-43]). When the velocity of the ionization front falls below this critical value, a shock wave appears and the ionization front is no longer R type. From equations (5-50) and (4-56) it follows directly [11] that dr_i/dt equals $2C_2$ when

$$\frac{1}{1 - r_i^3/r_S^3} = 1 + \frac{n_H \alpha^{(2)} r_i}{6 C_2} \qquad (5\text{-}52)$$

If we replace r_i on the right-hand side of equation (5-52) by r_S, equal to 9.5 pc for an O7 star with n_H equal to $10/\text{cm}^3$, and substitute from equation (4-59) for $\alpha^{(2)}$, with T equal to $10,000°\text{K}$, we find that the right-hand side of equation (5-52) equals 13. Hence the weak R-type front will persist until r_i is within about 3% of r_S.

Until r_i is nearly equal to r_S and S_u is substantially reduced by absorption, u_1 will much exceed C_2. Hence the velocity change Δu on passing through the ionization front will generally be much less than C_2, in accordance with equation (5-46). Throughout the ionized gas, velocities and corresponding perturbations will be present, but these are correspondingly small until r_i approaches r_S [11].

The detailed structure of the moving ionization front has been analyzed [10] by numerical means, including the equations of motion and continuity for the gas, the equation of radiative transfer for each ultraviolet frequency in the radiation field, the energy equation for the gas giving the rate of change of temperature resulting from compression and interaction with radiation, and the equation of ionization equilibrium. The diffuse ultraviolet flux was not considered in detail. For a star of surface temperature $42,000°\text{K}$ the results show a rapid temperature rise to nearly $30,000°$ in the ionization front, followed by a more gradual decline to the equilibrium temperature of somewhat less than $10,000°$; as noted in Section 4.4, this equilibrium T changes somewhat with position in the H II region.

Expansion of the Ionized Gas

The first phase in the development of an H II region, which has been discussed in the preceding paragraphs, is essentially the ionization of a large sphere of gas by the O star; motions are of small importance throughout most of this phase. In the second phase the heated sphere of gas tends to expand into the surrounding neutral region, where the density is presumably about the same as the initial H II density, but the pressure is less by some 2 orders of magnitude. Since the equilibrium radius r_S of the ionized region varies as $n_H^{-2/3}$, in accordance with equation (4-56), $n_H r_S^3$ varies as $1/n_H$, and hence the amount of material ionized increases as the gas expands. This is because with the same number of photons more hydrogen atoms can be kept ionized if n_H is reduced, since the rate of recombination per cm^3 varies as n_H^2. As a result, when the H II gas expands, the ionization front will move outwards through the gas. If the brightness of the O star were constant and the cloud of gas were sufficiently large, this second phase would terminate when pressure equilibrium was established. In fact the ultraviolet luminosity of the brighter stars will drop after a few million years.

The transition between these two phases is difficult to follow analytically. During this period the changes in velocity through the ionization front are comparable with the sound velocity in the ionized region, which we here denote by C_{II}. A shock forms at the ionization front and moves away, and the ionization front changes from weak R type at the beginning to weak D type at the end. This complex dynamical process can best be followed by direct numerical integration of the dynamical equations [10, 12]. While these computations do not include cooling of the neutral gas behind the shock front, they take most other features of the problem into account and show clearly the formation of the shock wave when the radius, r_i, of the ionization front approaches r_S, and dr_i/dt falls below $2C_2$.

After the shock has formed and has moved away from the ionization front, both fronts slow down somewhat and the ionization front becomes of weak D type and completely subsonic. The rate of expansion of the H II region during the ensuing second phase will now be deduced approximately from the known properties of the shock and ionization fronts.

As before, we denote by subscripts s and i quantities referring to the shock and the ionization fronts, with subscripts 1 and 2 referring to quantities in front of and behind these two fronts; ρ and v denote density and fluid velocity, respectively. Thus ρ_{s1} and ρ_{s2} are the densities ahead of and behind the shock front, while ρ_{i1} and ρ_{i2} are the densities ahead of and behind the ionization front. Let V_s and V_i denote the velocities of the two fronts relative to the star. These velocities, like the fluid velocity v, are measured positive in the direction of increasing r. If u_{s1} and u_{s2} represent fluid velocities relative to the shock front, and u_{i1} and u_{i2} corresponding velocities relative to the ionization front, then these velocities are positive in the inward direction and are related to the other velocities by the equations

$$-u_{s1} = 0 - V_s \qquad -u_{s2} = v_{s2} - V_s \qquad (5\text{-}53)$$

$$-u_{i1} = v_{i1} - V_i \qquad -u_{i2} = v_{i2} - V_i \qquad (5\text{-}54)$$

In the first of equations (5-53) we have assumed that v_{s1} vanishes; i.e., that the surrounding gas is initially at rest.

If we now combine equations (5-20) and (5-48) we obtain

$$\frac{\rho_{i2}}{\rho_{s1}} = \frac{V_s^2 \xi \rho_{i1}}{2 C_{\text{II}}^2 \rho_{s2}} \qquad (5\text{-}55)$$

where C_{II} is the sound velocity in the H II region behind the ionization front. Similarly, we may compute v_{i2} from the velocities v_{i1} and v_{s2}, using the fact that v_{s1} vanishes; if we let C_{I} be the sound velocity in neutral hydrogen (either in front of or behind the shock front) we find

$$(v_{i2} - V_i)\frac{C_{\text{I}}^2 \xi}{2 C_{\text{II}}^2} = \frac{\rho_{s2}}{\rho_{i1}}\left(V_s - \frac{C_{\text{I}}^2}{V_s}\right) - V_i \qquad (5\text{-}56)$$

To apply these results we shall assume that ρ_{i2}, the density behind the ionization front, equals ρ_{II}, the mean density throughout the H II region; we shall use ρ_{I} to denote ρ_{s1}, the density in the unperturbed H I region ahead of the shock. In accordance with this assumption, we assume that the velocity of the gas behind the

5. DYNAMICS OF THE INTERSTELLAR GAS

ionization front has a value corresponding to uniform expansion of the ionized gas in the H II region as a whole. If we follow a particular shell of ionized gas, of radius r, $r^3 \rho_{\text{II}}$ will be constant during the expansion, giving

$$\frac{1}{r}\frac{dr}{dt} = -\frac{1}{3\rho_{\text{II}}}\frac{d\rho_{\text{II}}}{dt} \tag{5-57}$$

At the boundary of the ionization zone, where $r = r_i$, dr/dt equals v_{i2}. The velocity V_i of the ionization front, on the other hand, is given by the condition that $r_i^3 \rho_{\text{II}}^2$ is constant [see equation (4-56)], provided that the luminosity of the central star is constant, and hence

$$\frac{1}{r_i}\frac{dr_i}{dt} = \frac{V_i}{r_i} = -\frac{2}{3\rho_{\text{II}}}\frac{d\rho_{\text{II}}}{dt} \tag{5-58}$$

If now we combine these results for v_{i2} and for V_i we find

$$v_{i2} = \tfrac{1}{2} V_i = u_{i2} \tag{5-59}$$

where the second of equations (5-54) has been used.

If we let ρ_{s2}/ρ_{i1} equal unity, since ρ should be nearly constant across the thin shell of compressed matter between the two fronts, we obtain, using equation (5-59) and eliminating ξ from equations (5-49) and (5-55)

$$V_s^2 = C_{\text{II}}^2 \frac{\rho_{\text{II}}/\rho_{\text{I}}}{1 - \rho_{\text{II}}/4\rho_{\text{I}}} \tag{5-60}$$

where the difference between V_s and V_i has been ignored. The value of this difference, obtained from equation (5-56), is given by

$$\frac{V_s - V_i}{V_s} = \frac{\rho_{\text{I}} C_{\text{I}}^2}{\rho_{\text{II}} C_{\text{II}}^2}\left(1 - \frac{\rho_{\text{II}}}{4\rho_{\text{I}}}\right)\left(1 - \frac{\rho_{\text{II}}}{2\rho_{\text{I}}}\right) \tag{5-61}$$

Since u_{s2}/u_{i1} equals about $2(\rho_{\text{I}}/\rho_{\text{II}})(V_s/V_i)$, and generally exceeds two, the mass in the compressed shell between the two fronts is greater than half the mass swept up by the shock.

If $\rho_{II}/4\rho_I$ is ignored, equation (5-60) may be integrated directly; this result becomes exact as ρ_{II}/ρ_I becomes small. Since ρ_{II} varies as $r_i^{-3/2}$, for constant stellar luminosity, we define r_{i0} by the relationship

$$\frac{\rho_{II}}{\rho_I} = \left(\frac{r_{i0}}{r_i}\right)^{3/2} \tag{5-62}$$

The radius r_{i0} is nearly equal to the initial value of r_S, the radius of the ionization front in radiative equilibrium. Since the difference between V_s and V_i is small until pressure equilibrium is approached, we may set V_s equal to dr_i/dt, and obtain from equation (5-60)

$$\frac{r_i}{r_{i0}} = \left(1 + \frac{7}{4}\frac{C_{II}\, t}{r_{i0}}\right)^{4/7} \tag{5-63}$$

The validity of the assumptions leading to equations (5-60) and (5-63) may be verified by examination of the numerical results obtained for old H II regions [12, 13]. In the numerical computations the full dynamical equations were integrated for times up to 2×10^6 years after the birth of an O7 star in a uniform H I region. The basic assumption of uniform expansion of the ionized gas, leading to equation (5-59), appears to be correct on the average, since the mean value of $2u_{i2}/V_i$ after the shock separates from the ionization front equals 1.00 in the computer results, with an average deviation of ± 0.05. For an initial H density, n_I, equal to 6.4 cm^{-3}, the value of $d \log r_i/d \log t$ in the computations falls to 0.55 for t equal to 2×10^6 years, at which time ρ_{II}/ρ_I has decreased to about 0.10; the theoretical asymptotic value of this slope should be 0.57, according to equation (5-63). Since the cooling of neutral hydrogen behind the shock front was not included in the numerical computations, the shell between the two fronts was found to be relatively thick, and equation (5-61) is not applicable. Because of the large thickness of this shell, ρ_{i1}/ρ_{s2} was less than unity, and the computed value of this quantity must be inserted into equation (5-60) to preserve agreement with the computer results. We conclude that equations (5-60) through (5-63) give a realistic description of an expanding H II region during its later stages.

The H I shells observed around H II regions tend to be rather thick, while the theory would predict thin shells until late in the expansion. As the ultraviolet luminosity of the O star decreases, the shell will tend to thicken, since the ionization front may even reverse its direction and move inwards, while the neutral shell will tend to continue its outwards expansion. The presence of a magnetic field will also thicken the shell of neutral gas between the shock and ionization fronts, by an amount depending on the field strength parallel to the shock front [see equation (5-23)].

Efficiency of Acceleration

The efficiency, ε_u, for acceleration of the interstellar gas by ultraviolet radiation may be defined as

$$\varepsilon_u = \frac{\tfrac{1}{2}Mv^2}{S_u(0)kT_c\psi_3(\beta_c)} \tag{5-64}$$

where M is the mass of the shell accelerated to a velocity v, $S_u(0)$ is the total number of photons beyond the Lyman limit radiated by the star per second, and $kT_c\psi_3(\beta_c)$ is the mean available energy per photon [see equation (4-78)]. Instead of computing ε_u directly from equation (5-64) we shall estimate this quantity from simple physical considerations.

During the isothermal expansion of a gas from density ρ_1 to a density ρ_2, the work done per particle equals $kT \ln \rho_1/\rho_2$, which equals kT when the density has decreased by a factor $1/e$, corresponding to an increase of radius of the ionization front by about a factor 2 [see equation (5-62)]. Per H nucleus the work done will be twice this, since the electrons contribute equally. The efficiency is the ratio of this work done to the power input during the time interval r_{i0}/C_{II}, during which r_i doubles; the radius, r_i, of the ionized region is assumed to be increasing at about the isothermal sound speed, C_{II} [see equation (5-63)]. Since the number of photons absorbed per H nucleus during this period is the effective number of recombinations per H nucleus, we have

$$\varepsilon_u = \frac{2kTC_{II}}{r_{i0}n_H \alpha^{(2)}kT_c\psi_3(\beta_c)} = \frac{T}{3T_c\psi_3(\beta_c)}\left[\frac{6C_{II}}{r_{i0}n_H\alpha^{(2)}}\right] \quad (5\text{-}65)$$

where we have inserted the same mean energy per absorbed photon used in equation (5-64), and replaced n_p by n_H. The quantity in brackets has already been discussed in connection with equation (5-52), and represents the drop in efficiency due to the many times that an electron is recaptured and photoejected as the density drops by $1/e$. The quantity $T/T_c\psi_3(\beta_c)$, which represents the drop in efficiency due to the reduction of the mean particle energy below the mean energy per absorbed photon, is about 1/5 for an O7 star, if T is set equal to 10,000°. Combining these results gives

$$\varepsilon_u \approx 0.006 \quad (5\text{-}66)$$

Since ε_u varies as $[S_u(0)n_H]^{-1/3}$, the efficiency will increase somewhat during the expansion; for the cooler stars also ε_u will be somewhat greater than the value in equation (5-66). Detailed computations of ε_u from equation (5-64), using M and v computed from equations (5-63) and (5-60) with an initial n_H of 10 cm^{-3}, yield [13] values ranging from 0.006 for a star of type O7 to 0.020 for one of type O9. Since the energy dissipation rate, P_c, given in equation (5-24) is about 0.014 times the available ultraviolet power, P_u, in equation (5-26), these computed efficiencies seem not inconsistent with the maintenance of cloud velocities by ultraviolet radiation.

5.4 Supernova Shells

The explosion of a supernova releases a vast amount of energy. Some of this will be radiated at a relatively early stage; some will be converted into kinetic energy of the expanding gas, which will sweep up, accelerate, and heat the surrounding interstellar material. Some gas may escape from the Galaxy entirely, especially if the supernova is far from the galactic plane. The gas that is slowed

5. DYNAMICS OF THE INTERSTELLAR GAS

down within the Galaxy will ultimately radiate most of its energy in some region of the electromagnetic spectrum. In this Section we attempt to follow theoretically the successive stages that may be expected following a supernova explosion. The discussion is necessarily confined to an idealized situation, in which the interstellar gas is homogeneous. Any numerical values computed will refer to a supernova of type I, for which, as we have seen in Section 5.1, a rough determination of the total initial kinetic energy, E, of the ejected envelope gives 10^{51} ergs [4].

Initial Expansion of a Supernova Atmosphere

When a supernova ejects a large mass of gas, the density near the star is much greater than the interstellar density, and the mass expands essentially into a vacuum. Analogy with terrestrial explosions suggests that a shock will begin to form when the material has swept over a distance about equal to the mean free path. If this were the correct criterion for material ejected from a supernova, a shock would never form, since the computed range is comparable with the dimensions of the Galaxy. The range of protons moving through neutral hydrogen atoms at a velocity of 2×10^9 cm/sec, corresponding to an energy of 2 MeV, may be computed from equation (4-29), which gives a value of 1.1×10^{-17} cm^2 for the ionization cross section, σ_{HR}. Since the energy lost per primary ion is about 60 eV, the energetic protons will lose all their energy after some 3×10^4 ionizing collisions, giving a range of roughly 1000 pc if the mean $n(\text{H I})$ is 0.85 cm^{-3}. Consideration of the increase in σ_{HR} as the proton slows down decreases the range to 700 pc, comparable with the mean free path of a visual photon in the galactic plane; the mean travel time is 1.4×10^{12} sec. Evidently a conventional shock cannot be formed around a type I supernova.

However, this simple picture must be modified by several effects. According to equation (5-8) the charged particles will gyrate around the magnetic field with a radius of gyration of only 10^{11} cm, for protons with $w_\perp$ equal to 20,000 km/sec in a field of 3×10^{-6} g. While the magnetic field has much too small an energy density to hinder the outwards expansion, it provides a massless barrier between the supernova gas and the ionized interstellar material; a hydromagnetic shock will form, and move outwards. Moreover, as the gas expands, the magnetic field may weaken, in which case the kinetic energy of gyration around the lines of force will decrease adiabatically. Thus the mean energy of the particles initially in the envelope may decrease as the shock sweeps up more and more material.

The particle motion parallel to **B** is unaffected by the magnetic field. Thus one might expect that one-third of the kinetic energy would be unaffected by phenomena in the neighborhood of the supernova, and would appear in the form of energetic particles several hundred parsecs away. However, there are a number of additional mechanisms by which particles can lose energy even if their initial velocity is parallel to **B**. Penetration of the positive ions through the surrounding interstellar gas may excite plasma oscillations [14], which under some conditions can slow down the ions. In addition, if a supernova explosion pushes the field lines back and creates a cavity in the field [15], it is not obvious how energetic particles can escape from this cavity without retardation and end up by spiralling along the lines of force of the interstellar field.

Evidently the fraction of the supernova energy that appears far away from the explosion in the form of 2 MeV protons (together with He^{++} nuclei with similar energies per nucleon) is not likely to exceed one-third and may be much less. We shall take one-third to give an upper limit; from equation (5-27) the energy available from such particles is 6×10^{-26} ergs cm^{-3} sec^{-1}, the value used in equation (4-45) for ζ_s, the upper limit on the ionization rate by energetic particles. The energy density, u_{Rs}, for these energetic particles emanating from supernova, is roughly equal to the energy dissipation rate times the travel time of 1.4×10^{12} sec for each particle, giving

$$u_{Rs} \approx 6 \times 10^{-26} \times 1.4 \times 10^{12} \approx 8 \times 10^{-14} \text{ erg/cm}^3 \quad (5\text{-}67)$$

less than one-tenth the energy density of cosmic rays observed at the earth [see equation (2-66)]. Thus the energy density of these

5. DYNAMICS OF THE INTERSTELLAR GAS

energetic particles is relatively unimportant even though, as shown in the preceding chapter, they may play such a major role in ionizing and heating H I regions.

Even if as much as one-third of the supernova energy appears in the form of freely moving relativistic particles, it seems clear that the remaining fraction will be communicated to the interstellar medium through a hydromagnetic shock, moving across the interstellar magnetic field. At least two-thirds of the initial energy of the supernova shell is probably transformed in this way, and perhaps nearly all. To follow the evolution of this energy we shall adopt the approximation of a spherical shock and shall ignore the magnetic field; despite the obvious physical shortcomings of this picture, the results which it yields may reproduce in a general way the physical effects associated with the actual situation. The remainder of the section is devoted to an analysis of this idealized spherical expansion of a supernova envelope.

Three different stages in the expansion may be distinguished. In the first stage the interstellar material has little effect because of its low density, and the velocity of expansion of the supernova envelope will remain nearly constant with time. This phase terminates when the mass of gas swept up by the outwards moving shock equals the initial mass, M_s, expelled from the supernova; i.e., when

$$\frac{4\pi r_s{}^3 \rho_1}{3} = M_s \tag{5-68}$$

where ρ_1 is the density of the gas in front of the shock, and r_s is the radius of the shock front. For M_s equal to 0.25 $M_\odot$ and ρ_1 equal to 2×10^{-24} g/cm^3, equation (5-68) will be satisfied when r_s equals 1.3 pc, which will occur about 60 years after the initial explosion.

During the second phase of expansion the mass behind the shock is determined primarily by the amount of interstellar gas swept up, but the energy of this gas will be constant; radiation during this period is unimportant. This phase is discussed immediately below. In the third phase, which is treated subsequently, radiation becomes dominant and the shock becomes effectively isothermal.

Intermediate Nonradiative Expansion

When the interstellar mass swept up by the outgoing shock exceeds the initial mass, the shock velocity will begin to drop. The temperature of the gas is so high at this stage that the radiation rate will be small and will be neglected here to give a first approximation. More specifically, the electrons will be heated initially to about 2×10^7 deg on passage through a shock at 20,000 km/sec, and interaction with the positive ions will raise the electron temperature even higher. Thus we may assume that the total energy of the gas within the shock front is constant, and equal to the initial energy, E.

While a detailed determination [16] of the motions in this case is somewhat involved, an equation for the outwards velocity of the shock can be computed readily from simple physical principles. Let a fraction K_1 of the total energy E be in heat energy, with the balance in kinetic energy, and let p_2, the pressure immediately behind the shock front, equal K_2 times the mean pressure of the heated gas within the spherical shock. If the material is assumed to be a perfect gas, with γ equal 5/3, then the mean pressure is 2/3 the mean density of thermal energy, giving

$$p_2 = \frac{3K_2}{4\pi r_s^3} \frac{2K_1 E}{3} = \frac{KE}{2\pi r_s^3} \quad (5\text{-}69)$$

where K equals $K_1 K_2$, and r_s is again the radius of the shock. In the shock condition (5-18), p_1 may be neglected, since the shock is strong, giving immediately for the shock velocity,

$$u_1^2 = V_s^2 = \frac{2KE}{3\pi \rho_1 r_s^3} \quad (5\text{-}70)$$

According to the detailed dynamical theory, a similarity solution exists for which the structure of the expanding shell is constant with time. From the exact computations for this solution, in the case where γ equals 5/3, the numerical values of K_1 and K_2 obtained are 0.8 and 2, respectively [16], and hence K equals 1.6. The exact solution also indicates that most of the mass is in the outer 20% of the radius of the expanding sphere.

5. DYNAMICS OF THE INTERSTELLAR GAS

The temperature immediately behind the shock front may be obtained from p_2/ρ_2, using the usual formula for a perfect gas. In accordance with equation (5-19), ρ_2 equals $4\rho_1$, the usual result for a strong shock with γ equal to 5/3, and from equation (5-18) we obtain for T_2

$$T_2 = \frac{3\mu}{16k} V_s^2 = \frac{0.064\mu}{k} \frac{E}{\rho_1 r_s^3} \qquad (5\text{-}71)$$

where K has been replaced by its value 1.6, and where μ is the mean mass per particle, equal to 0.61 m_H for the helium–hydrogen ratio of 0.10 given in Table 4.7, if full equipartition of thermal energy is assumed between electrons and ions; such equipartition is certainly not valid at the outset, but should become so as the shock gradually slows down.

Equation (5-70) may be integrated at once, since V_s equals dr_s/dt and we have

$$r_s = \left(\frac{2.1E}{\rho_1}\right)^{1/5} t^{2/5} = \frac{0.32 t(\text{years})^{2/5}}{n_H^{1/5}} \text{ parsecs} \qquad (5\text{-}72)$$

where the numerical value has been computed for E equal to 10^{51} ergs, and ρ_1/n_H has been computed with n_{He}/n_H again equal to 0.10. If we substitute this result in equation (5-71), we obtain*

$$T_2 = \frac{0.03\mu}{k}\left(\frac{2.1E}{\rho_1}\right)^{2/5} t^{-6/5} = \frac{2.2 \times 10^{11}}{t(\text{years})^{6/5} n_H^{2/5}} \text{ degrees K} \qquad (5\text{-}73)$$

We have seen that this type of motion is applicable only for t greater than a few hundred years, somewhat after equation (5-68) is first satisfied. In addition, these results are valid only as long as radiative cooling of the gas remains unimportant. When T falls much below 10^6 deg, corresponding to mean electron energies of about 100 eV, the abundant ions C, N, and O will begin to acquire bound electrons, and excitation of these will radiate energy

---

* An earlier version of equation (5-73), which has been widely used [4,17], gives T_2 too great by a factor 4; if ion-electron equipartition is assumed, these earlier values of T_2 should be divided by 8.

rapidly [17]. At these temperatures electrons and ions will be close to equipartition, since the equipartition time, $t_s/2$, found from equation (4-12) becomes relatively short—about $400/n_p$ years. We may assume that radiation cooling becomes dominant for T somewhere between 10^6 and 10^5 deg, corresponding to values of t between 2×10^4 and 2×10^5 years, and to radii between 10 and 20 pc, again computed for n_H equal to unity. If n_H is as low as 0.1, radiative cooling should become dominant for radii about twice this large.

Late Isothermal Expansion

When radiative cooling becomes important, the temperature of the gas will fall to a relatively low value. The outwards progress of the shock is now maintained not by the stored thermal energy, which is rapidly radiated, but by the momentum of the outwards moving gases. The shock itself may be regarded as an isothermal shock. According to equation (5-20), the compression across the shock will be very large, and from equation (5-15) we see that the velocity u_2 of the gas away from the front is very slow. Material that has crossed the front will stay close to the front for a long time, forming a thin compressed shell of outward moving gas. The thickness of the shell will be determined primarily by the cooling time for gas that has been heated in the initial nonradiating shock (see Fig. 5.1).

The velocity of the shell during this phase may be computed from the condition that the outwards momentum remains constant. This simple picture is sometimes referred to as the "snowplow model," since the interstellar gas is swept up by the expanding shock in somewhat the same manner as snow accumulates on the front of a snowplow.

If M_t and v_t are the mass and velocity of the shell when this model first becomes applicable—when the shell is in transition from the nonradiative to the radiative model—we have

$$Mv = M_t v_t \qquad (5\text{-}74)$$

5. DYNAMICS OF THE INTERSTELLAR GAS

By the time equation (5-74) is applicable, the mass in the shell will be largely interstellar material swept up in an earlier phase. Hence M varies as r_s^3, where r_s is the shell radius, and an integration yields

$$t = \frac{\pi \rho_1}{3 M_t v_t} r_s^4 = \frac{M r_s}{4 M_t v_t} \tag{5-75}$$

The retardation of the shell is now more rapid than before, with dr_s/dt varying as $1/r_s^3$, as compared with the $1/r_s^{3/2}$ variation given in equation (5-70).

Most observed supernova remnants are smaller than the radii at which the snowplow model becomes applicable, although these results may be applicable to the Cygnus Loop [4]. However, if the energy required for maintaining interstellar motions comes from expanding supernova shells, this model may be used for computing ε_s, the efficiency with which supernova energy is converted into cloud motions. If a supernova shell finally produces a total mass, M_c, moving at the velocity v_c, which we identify with the r.m.s. cloud velocity of 14 km/sec, then ε_s is given by

$$\varepsilon_s = \frac{M_c v_c^2}{2E} = \frac{M_c v_c^2}{M_t v_t^2} \times \frac{M_t v_t^2}{2E} \tag{5-76}$$

where M_t and v_t are again the mass and velocity at which the shock becomes isothermal instead of nonradiating. We have already seen that behind the nonradiating shock the kinetic energy is about one-fifth of the total energy. If we use equation (5-74) we obtain

$$\varepsilon_s = \frac{v_c}{5 v_t} \tag{5-77}$$

To maximize ε_s we take a minimum value for v_t, and assume that the transition between the two models occurs when equation (5-71) predicts a temperature of 10^5 deg K. Then v_t equals 85 km/sec, and

$$\varepsilon_s = 0.033 \tag{5-78}$$

According to equations (5-24) and (5-27) the observed P_c is 0.03

times P_s and the required efficiency is about equal to the value of ε_s in equation (5-78). If v_t were taken to correspond to a higher temperature, however, this agreement would disappear. Evidently supernovae play an important role in interstellar gas dynamics, and may provide the major energy source for the observed motions of the interstellar gas.

All the analyses in this section have been based on a uniform density of the gas in front of the shock. In fact the interstellar gas seems to be quite inhomogeneous; the propagation of strong shocks through regions containing density irregularities has not been extensively studied. In regions where the density is higher than the average, one would expect the shock front to move more slowly and the resultant rise of temperature to be less. This effect may explain in part why rather normal spectra are observed from supernova remnants, although the velocity is so great that complete ionization might be expected.

5.5 Interaction between Clouds and Stars

There are many sorts of dynamical interactions between interstellar clouds and individual stars. In this section we shall consider processes which are of importance mainly for the interstellar material, ignoring those interactions which are of interest primarily because of their effect on the stars. Thus gravitational interactions between stars and cloud complexes, which may alter the velocity distribution of the stars, and accretion of interstellar gas by slowly moving stars, which may increase the masses of such stars, will both be ignored. Instead we consider the changes produced in the velocity and physical state of an interstellar cloud when it encounters the ionizing radiation of an early-type star, or when it is subject to the radiation pressure produced by any star of high luminosity.

Ionization of an H I Cloud in an H II Region

A dense cloud of neutral hydrogen may be present within an H II region either because a cloud of relatively higher density was

5. DYNAMICS OF THE INTERSTELLAR GAS

present when the early-type star was born, or because an H I cloud has moved into the region. In either case an ionization front will gradually eat its way into the cloud. This phenomenon is of interest partly because it presumably produces the bright rims observed in H II regions (see frontispiece) and partly because it may also produce a large acceleration of the cloud.

To analyze this situation we consider a spherical cloud, sufficiently close to the central star so that the self-absorption of ultraviolet radiation in the H II region may be neglected. The ionization front will be preceded by a shock, the two moving into the cloud at about the same velocity, determined by equation (5-40) and the quantity J, defined as the number of ultraviolet photons reaching the ionization front per cm^2 per sec.

While the situation is similar to the general expansion of an H II region, the geometry in detail is very different. In particular, the ionized gas immediately behind the ionization front will generally be at a much higher density than the surrounding ionized gas, and will expand spherically, going away from the cloud at a velocity somewhat exceeding the local velocity of the sound. As a result, the density ρ_{i2} and the velocity u_{i2} immediately behind the front are not separately determined, as they were in Section 5.3 [see equations (5-62) and (5-59)]. In the absence of any complete dynamical discussion, previous analyses of the problem [18, 19] have generally made two simplifying assumptions. In the first place, the ionization front is assumed D-critical; this assumption is consistent with evidence on the similar detonation fronts, which in accordance with the Chapman-Jouguet hypothesis tends to be D-critical when not constrained by boundary conditions to be of some other type [1]. From this assumption it follows that u_{i2} equals C_{II}, and the density change across the front is given by equation (5-47). In the second place, the velocity of the ionized gas as it streams away from the front is assumed constant and equal to C_{II}; this assumption simplifies the analysis, but certainly underestimates the gas velocity, which must be increased by the negative pressure gradient. Errors in these assumptions are not likely to have a major qualitative effect on the results.

The ionized atoms which are streaming away from the cloud recombine with electrons in part to form an "insulating layer," which absorbs some of the incoming ultraviolet radiation and may seriously reduce the number of such photons reaching the ionization front. Let J_0 be the particle flux of photons in the absence of this layer; evidently

$$J_0 = \frac{S_u}{4\pi r^2} \tag{5-79}$$

where r is the distance from the central star, and S_u has the same meaning as $S_u(0)$ in Section 4.3. As in Section 4.3 we equate the rate of absorption to the rate of ion-electron recombination, which varies as n_i^2. To determine n_i we use the equation of material continuity in the form

$$n_i(R) = \frac{J}{C_{\text{II}}} \left(\frac{R_i}{R}\right)^2 \tag{5-80}$$

where R is the distance from the cloud center, and R_i is the value of R at the ionization front. In equation (5-80) the velocity of the ionization front through the compressed material in the cloud has been assumed small, and as a result C_{II} is the velocity of the ionized gas relative to the cloud as well as relative to the ionization front.

The difference between J_0 and J is simply the number of recombinations in a column 1 cm^2 in cross section extending through the insulating layer towards the central star. Hence we have

$$J_0 - J = \int_{R_i}^{\infty} \alpha n_i^2(R) \, dR = \frac{\alpha J^2 R_i}{3 C_{\text{II}}^2} \tag{5-81}$$

where the integration has been carried out after substituting equation (5-80) for n_i. In equation (5-81) α will be intermediate between $\alpha^{(1)}$ and $\alpha^{(2)}$, since the amount of diffuse ultraviolet radiation which escapes from the cloud will vary with the thickness of the insulating layer. While equation (5-81) is strictly valid only along the line from the cloud center to the star, we shall obtain approximate results by applying this equation to the entire hemisphere illuminated by the central star.

5. DYNAMICS OF THE INTERSTELLAR GAS

Equation (5-81) is quadratic in J_0/J, yielding the solution

$$\frac{2J_0}{J} = 1 + \left\{1 + \frac{\alpha S_u R_i}{3\pi r^2 C_{\text{II}}^2}\right\}^{1/2} \tag{5-82}$$

where equation (5-79) has been used to eliminate J_0. A typical value of R_i is about 0.2 pc, and if this value is inserted into equation (5-82), with S_u taken from Table 4.5, α equated to 4.1×10^{-13} sec^{-1} cm^3, and C_{II} set equal to 11 km/sec, J_0/J equals 8 for a cloud 5 pc from on O7 star. The corresponding value of n_i immediately behind the ionization front is about 10^2 cm^{-3}, and the density of neutral atoms in the cloud immediately ahead of the ionization front must be about 10^5 cm^{-3} [see equation (5-47)]. A preceding shock presumably compresses the gas to this high density.

One may reasonably identify the observed bright rims in H II regions with the luminous insulating layers of ionized gas discussed here; the observed radius of curvature of a bright rim may be identified with R_i. The measured velocity of approach of the emitting atoms, noted in Section 2.2, amounts to 13 km/sec, in reasonable agreement with theoretical expectations, based on C_{II} equal to 11 km/sec. The observed rim thickness averages between 15 and 20% of the observed R_i; this result is not inconsistent with a surface brightness varying as the inverse cube of the projected distance from the center, for distances exceeding R_i; this variation follows when n_i^2 in equation (5-80) is integrated along a line of sight. A detailed study of these rims would have to take into account the change of temperature with position in the insulating layer, as has been done for the gas behind the ionization front in an expanding H II region [10].

Acceleration of an H I Cloud by the Rocket Effect

The ionized gas ejected from an H I cloud in an H II region will mostly be traveling towards the central star. The inwards momentum carried away per second by this stream must just equal the outwards momentum gained by the cloud. Exactly as in an ordinary rocket the change of v, the mean velocity of the cloud, whose mass equals M, is given by

$$M\frac{dv}{dt} = -V\frac{dM}{dt} \tag{5-83}$$

where $-dM/dt$ is the mass carried away per second by the stream of ionized gas, and V is the inward component of the mean velocity of the ejected gas relative to the cloud. If M_0 and v_0 are the initial mass and velocity of the cloud, equation (5-83) yields the solution

$$M = M_0 e^{-(v-v_0)/V} \tag{5-84}$$

If a large fraction of the cloud mass is lost by ionization and ejection, the change in velocity of the remaining mass of cloud can substantially exceed V, which, in turn, is likely to be somewhat greater than the isothermal sound speed C_{II}.

The amount of mass that can be evaporated from a cloud is limited, in turn, by the acceleration, which drives the cloud further from the star and reduces dM/dt. To compute the mass of a cloud that can be almost completely ionized, we let

$$\frac{dM}{dt} = -\pi R_i^2 J \mu_i \tag{5-85}$$

where μ_i is the mean mass of gas per atom ionized; as noted in Section 5.3 μ_i is 1.40 m_{H} if helium remains neutral, and 1.27 m_{H} if the helium becomes singly ionized. If we combine equations (5-85), (5-79), and (5-82), neglecting terms of order J/J_0 in this last equation, and substitute into equation (5-83), together with (5-84) also, substitution of $v\,d/dr$ for d/dt and integration yield

$$\left(1 + \frac{v}{V}\right)e^{-v/V} = 1 - \frac{\mu_i R_i^{3/2}}{2M_0}\left(\frac{3\pi S_u}{\alpha}\right)^{1/2} \ln \frac{r}{r_0} \tag{5-86}$$

where the velocity has been assumed zero at r equal to r_0, and C_{II} has been equated to V. The cross section of the cloud has been assumed constant with time.

If appropriate numerical values are inserted, with r/r_0 equal to 10, and S_u again set equal to 2.7×10^{48} sec^{-1}, the value given for an

5. DYNAMICS OF THE INTERSTELLAR GAS

O7 star in Table 4.5, we find that for R_i between 0.2 and 10 pc the right-hand side of equation (5-86) vanishes for M_0 between about 5 and 2000 $M_\odot$. The higher values are not unrepresentative for interstellar clouds; however, the condition that the ionization front not be R type requires that these larger, less dense clouds be accelerated only rather far away from the central star.

While this mechanism of acceleration seems clear enough in principle, a detailed theory is needed to clarify a number of difficult points. In particular, passage of a single shock will accelerate material only up to about the sound velocity, C_{II}. Thus continuing acceleration of a cloud to high velocities would seem to require many successive shocks. One might expect that the shock front which starts at the inner surface of the cloud would travel to the outer surface and there generate a rarefaction wave that returns to the inner surface, the process being repeated several times. It is not clear whether lateral expansion of the compressed material during this time would lead to disruption of the cloud.

Radiation Pressure on Grains near a Bright Star

A solid particle near a star will be subject to an outward force of radiation pressure that even near the Sun may substantially exceed the inwards force of gravity. Near the more luminous stars this force can be relatively much larger. If L is the luminosity of the star, the force of radiation pressure on the grain, denoted by F_r, varies with distance, r, from the star according to the relation

$$F_r = \frac{La^2 Q_p}{4cr^2} \quad (5\text{-}87)$$

where Q_p is a mean value of $Q_p(v)$, the efficiency factor for radiation pressure, averaged over the spectrum of the star. For a dielectric grain Q_p may be substantially less than Q_e, since the radiation is scattered predominantly in a forward direction. For an index of refraction equal to 1.33 and $2\pi a/\lambda$ equal to 4, corresponding to a radius of 2.9×10^{-5} cm (see Section 3.2) and a wavelength of 4500 Å, $Q_p(v)$ equals 0.7 [20], as compared with a value of 2.8 for

$Q_s(v)$ (see Fig. 3.1). Near an early-type star the effective wavelength is less, tending to decrease Q_p, but the absorption is presumably greater (imaginary part of m larger), tending to increase Q_p. We shall here set Q_p equal to 0.7.

The velocity attained by the grain under the influence of this force will be limited by interactions between the grain on the one hand and the magnetic field and the interstellar gas on the other. The gas will in general not move across the magnetic field, and the grains will under some conditions tend to be motionless with respect both to the gas and the magnetic field.

Interactions with the magnetic field will be considered first. Because of the electric charge on the grain, these particles will tend to gyrate around the magnetic lines of force with an angular velocity, ω_B, given by equation (5-7). If we define the time t_B as equal to $1/\omega_B$, then for a grain of radius 2.9×10^{-5} cm (see Section 3.2) we obtain in a field 3×10^{-6} G

$$t_B = \frac{4\pi a^3 \rho_s c}{3 Z_g e B} = \frac{6.8 \times 10^4}{Z_g} \text{ years} \qquad (5\text{-}88)$$

where we have set ρ_s, the density of solid matter within the grain, equal to 1 g cm^{-3} in computing the mass of the grain. In an H II region, where Z_g equals about 450, t_B is very short. Even in an H I region, t_B is usually shorter than the dynamical times of interest, which are generally 10^6 years or more. A single grain in an H I region will spend a certain fraction of its time with no charge at all, but since the mean time between electron impact is at most 10^5 sec at the average density of one atom per cm^3, and a minimum electron density of 5×10^{-4} cm^{-3}, 10^5 sec represents a typical limit for the time during which a grain can travel across the magnetic field. We conclude that under most conditions the motion of a grain relative to the gas is limited to a direction parallel to **B**.

For such parallel motions, **B** may be ignored, and collisions between grains and atoms will limit w, the velocity of a grain relative to the gas. These collisions will produce an exchange of momentum given by $m_g w/t_s$, where m_g is the grain mass and t_s is the "slowing-down time" discussed in Section 4.1. If we use

5. DYNAMICS OF THE INTERSTELLAR GAS

equation (5-87), consider only radial motions directly towards or away from the star, and ignore the magnetic force, the equation of motion for a grain becomes

$$\frac{dw}{dt} = \frac{La^2 Q_p}{4cr^2 m_g} - \frac{w}{t_s} \qquad (5\text{-}89)$$

If the second term on the right-hand side, giving the collisional drag, is ignored and the grain is initially at a distance r_0 from the star, w will increase to an asymptotic value, w_a, given by

$$w_a{}^2 = \frac{3LQ_p}{8\pi c r_0 a \rho_s} \qquad (5\text{-}90)$$

where m_g, the grain mass, has again been replaced by $4\pi a^3 \rho_s/3$. For example if a grain 2.9×10^{-5} cm in radius, and with ρ_s equal to unity, is initially 1 pc from a star of luminosity $10^5 L_\odot$, then w_a equals 34 km/sec.

If the drag term is sufficiently large, w will reach a steady value, w_D, called the "drift velocity," at which the outwards force equals the drag. At the radius r_0 this drift velocity is given by

$$\frac{w_D}{w_a} = \frac{w_a t_s}{2r_0} \qquad (5\text{-}91)$$

Equation (5-91) is valid only if $w_a t_s/r_0$ is small compared to unity. In an H II region t_s may be computed from equation (4-12), with protons and He$^+$ ions as the field particles. If Z_g, the charge on the grain, is obtained from equation (4-96) we find

$$t_s = 0.17 \frac{a\rho_s}{n_H (kTm_H)^{1/2} \ln \Lambda} \qquad (5\text{-}92)$$

where again n_{He}/n_H has been set equal to 0.10. For an H II region with n_H equal to 10, t_s is about 500 years for the same grain parameters used here; with r_0 and w_a equal to 1 pc and 20 km/sec, respectively, $w_a t_s/r_0$ in equation (5-91) is about 10^{-2}. Hence w_D in this case is only 0.1 km/sec, and during 10^6 years the grains will move only 0.1 pc with respect to the interstellar gas. It will be seen from equations (5-90) through (5-92) that w_D is independent of a and ρ_s.

In an H I region collisions with neutral hydrogen and helium atoms provide the drag, and t_s must now be obtained from equation (4-9). With the usual formulae used for the cross section and mass of the grain, and with the same helium–hydrogen ratio assumed as before, we find

$$t_s = 0.70 \frac{a\rho_s}{n_H(kTm_H)^{1/2}} \tag{5-93}$$

With T equal to 50°, but with other quantities the same as before, t_s equals 6×10^5 years, and $w_a t_s/r_0$ is 12, indicating that under these conditions the drag is relatively unimportant, and the grains will be accelerated to the velocity w_a, slowing down gradually at much greater distances from the star.

On the assumption that the grains are dielectric particles we may conclude that near an early-type star, where the hydrogen is ionized, grains and gas are generally coupled together dynamically. Close to a late-type supergiant, where an H I region is to be expected, the grains may separate from the gas in a direction parallel to the magnetic field.

One important exception to this conclusion is the case of a supernova, for which the value of L in visible light is about 10^{10} $L_\odot$ [4], giving a radiative force much greater than the drag for particles within a few parsecs. During the short lifetime of a supernova, a sufficiently small grain may reach a velocity exceeding 10^4 km/sec, though the total energy input into such high-speed grains would appear to be small [21].

Radiation Pressure of Galactic Light on Grains

The general galactic light will produce a small force of radiation pressure on grains. This force is not likely to produce much effect in H I clouds, where t_s has the value 6×10^5 years found above. In regions of lower density, however, radiation pressure from this source may be important. If n_H is 0.25 per cm^3 between the clouds

5. DYNAMICS OF THE INTERSTELLAR GAS

with T equal to $50°$, t_s rises to 2×10^7 years; even if the temperature of this rarefied medium is as great as $2000°$ (see Section 5.2) t_s equals 3×10^6 years. In either case, significant separation of gas and dust may be produced parallel to the magnetic field.

Under these conditions the radiation pressure on the grains will produce a force between grains, tending to push them together. If two identical grains are at a distance r from each other in an isotropic radiation field of energy density U_v, the absorption of radiation by each will weaken the radiation density at the other by an amount $\pi a^2 U_v Q_a / 4\pi r^2$, where Q_a is the efficiency factor for absorption at the frequency v. Scattering produces no effect if the radiation field is isotropic. Since the reduced energy density refers to photons all traveling along the line joining the two grains, the resultant force, F_r, on each grain, directed toward the other, is given by

$$F_r = \frac{\pi Q_a Q_p a^4 U}{4r^2} \tag{5-94}$$

where U is the total energy density and where a mean value of $Q_a Q_p$ has been taken over the radiation field; the efficiency factor, Q_p, for radiation pressure has been discussed above in connection with equation (5-87). The energy absorbed by each grain will be reemitted at a lower frequency, for which we assume that $Q_p(v)$ is negligibly small. The nonisotropic character of the galactic radiation will give a substantially greater force of radiation attraction in the galactic plane than is given in equation (5-94). Perpendicular to this plane the high albedo of the grains may produce a repulsive force between the grains.

The ratio of F_r in equation (5-94) to the gravitational force F_G between the two grains is given by

$$\frac{F_r}{F_G} = \frac{9 Q_a Q_p U}{64\pi G a^2 \rho_s^2} \tag{5-95}$$

If we assume that $Q_a Q_p$ equals 0.2, corresponding to Q_a/Q_e equal to 0.1 and Q_p equal to 0.7 as discussed above, take U equal to

7×10^{-13} erg/cm^3, and use the grain parameters adopted above, F_r/F_G equals 110. Between two volume elements of the interstellar gas this force will be relatively small compared to the gravitational attraction between the predominant hydrogen atoms, but this radiative attraction can conceivably have an important differential effect on the grains relative to the gas. For example, a grain at 20 pc from a standard cloud will experience a radiative force, resulting from absorption of galactic light by the cloud, which will produce a velocity of about 0.2 km/sec in 5×10^7 years, giving a distance traveled by the grain during this time of roughly 5 pc. However, the effective acceleration time may be 1 order of magnitude smaller than this assumed value. For example, if the gas in which the grains are embedded has a velocity of 10 km/sec relative to the cloud, the effective acceleration time is reduced to about 10^7 years, and, as we have seen in the preceding paragraph, t_s may be equally short if the temperature of the intercloud gas is relatively high. A tenfold reduction of the effective acceleration time would reduce the distance traveled by the grain through the gas by 2 orders of magnitude, down to a value of about 0.05 pc in 5×10^6 years. So small a separation would be quite unimportant. Uncertainties in $Q_a Q_p$ and in the intercloud density and temperature leave open the possibility that under some conditions this radiative force may have important consequences.

References

1. S. A. Kaplan, *Interstellar Gas Dynamics*, 2nd rev. ed. (edited by F. D. Kahn), Pergamon Press, New York, 1966
2. L. Spitzer, *Physics of Fully Ionized Gases*, Interscience, New York, 2nd rev. ed., 1962.
3. L. Woltjer, in *Stars and Stellar Systems*, Vol. 5, Univ. of Chicago Press. 1965, p. 531.
4. R. Minkowski, in *Stars and Stellar Systems*, Vol. 7, Univ. of Chicago Press, 1968, p. 623.
5. P. Katgert and J. H. Oort, *B.A.N.*, **19**, 239 (1967).
6. L. Spitzer, *Astrophys. J.*, **95**, 329 (1942).
7. L. Biermann and L. Davis, *Z. Astrophys.*, **51**, 19 (1960).
8. E. N. Parker, *Astrophys, J.*, **145**, 811 (1966).

5. DYNAMICS OF THE INTERSTELLAR GAS

9. F. D. Kahn and J. E. Dyson, *Ann. Rev. Astron. Astrophys.*, **3**, 47 (1965).
10. W. G. Mathews, *Astrophys. J.*, **142**, 1120 (1965).
11. P. O. Vandervoort, *Astrophys. J.*, **139**, 889 (1964).
12. B. M. Lasker, *Astrophys. J.*, **143**, 700 (1966).
13. B. M. Lasker, *Astrophys. J.*, **149**, 23 (1967).
14. P. D. Noerdlinger, *Astrophys. J.*, **133**, 1034 (1961).
15. R. Kulsrud, I. B. Bernstein, M. Kruskal, J. Fanucci, and N. Ness, *Astrophys. J.*, **142**, 491 (1965).
16. L. I. Sedov, *Similarity and Dimensional Methods in Mechanics* (translated from the Russian by M. Friedman), Academic Press, New York, 1959.
17. C. Heiles, *Astrophys. J.*, **140**, 470 (1964).
18. J. H. Oort and L. Spitzer, *Astrophys. J.*, **121**, 6 (1955).
19. S. R. Pottasch, *B.A.N.*, **14**, 29 (1958), No. 482.
20. H. C. van de Hulst, *Rech. Astron. Obs. Utrecht*, **11**, Part 1 (1946); esp. Fig. 23.
21. B. Wolfe, P. McR. Routly, A. S. Wightman, and L. Spitzer, *Phys. Rev.* **79**, 1020 (1950).

Chapter 6

Formation of Stars

6.1 Gravitational Instability

Star formation seems to be a general phenomenon occurring within our own Galaxy and in the spiral arms of similar galaxies [1]. The apparent concentration of young early-type stars in interstellar clouds of gas and dust can best be explained by the assumption that the interstellar medium is the birthplace of these new stars. Moreover, the observations suggest that new stars are primarily formed in clusters, though the isolated formation of a single star from an individual interstellar cloud cannot be excluded observationally.

The detailed steps by which stars might form from the interstellar gas cannot be outlined very precisely at the present time, since there are too many uncertainties. However, different physical processes that may occur can be analyzed and some of the different possibilities evaluated. A fundamental first step in star formation is the pulling together, by gravitational forces, of some large mass of gas to form a single condensation that is gravitationally bound. The present section discusses this first step, with the following sections treating subsequent evolutionary stages. Throughout all this discussion, analysis is possible only with idealized models, which generally provide only an approximate representation of reality.

6. FORMATION OF STARS

Linearized Perturbation Equations

The general concept of gravitational instability starts with a medium in equilibrium. A small disturbance is assumed to occur. Under some conditions the gravitational force will produce an exponential growth of the disturbance, with the density increasing in some regions and decreasing in others. The regions of increasing density represent the first stage in the development of gravitational condensations. The initial growth of such condensation can be discussed by a linearized theory, in which all changes from the equilibrium state are assumed small, and terms of second or higher order in these perturbations can be ignored. The later stages of growth, when the perturbations are appreciable, must be analyzed by a "finite-amplitude" theory, which is generally much more difficult.

In the initial equilibrium state the quantities p, ρ, $\mathbf{v}$, ϕ, and $\mathbf{B}$, which appear in equations (5-1) through (5-6), are all independent of time; we denote these equilibrium values by subscripts zero. The deviations from these equilibrium values will be denoted by subscripts 1. Hence we have, for example,

$$\rho = \rho_0 + \rho_1 \qquad (6\text{-}1)$$

To derive the first-order, or linearized, momentum equation for these deviations, we substitute equation (6-1) and similar relations for the other physical quantities in the momentum equation (5-1). The equilibrium quantities satisfy equation (5-1) separately and cancel out. If terms of second and higher order in ρ_1/ρ_0 and v_1/v_0 are neglected, one obtains

$$\frac{\partial \mathbf{v}_1}{\partial t} + \mathbf{v}_0 \cdot \nabla \mathbf{v}_1 + \mathbf{v}_1 \cdot \nabla \mathbf{v}_0 = -\nabla \left(\phi_1 + C^2 \frac{\rho_1}{\rho_0} \right) \qquad (6\text{-}2)$$

where we have assumed that the temperature of the gas is everywhere constant and that no magnetic field is present. Similarly, the continuity equation (5-2) becomes

$$\frac{\partial \rho_1}{\partial t} + \mathbf{v}_0 \cdot \nabla \rho_1 + \mathbf{v}_1 \cdot \nabla \rho_0 = -\rho_1 \nabla \cdot \mathbf{v}_0 - \rho_0 \nabla \cdot \mathbf{v}_1. \qquad (6\text{-}3)$$

while Poisson's law, equation (5-3), yields

$$\nabla^2 \phi_1 = 4\pi G \rho_1 \tag{6-4}$$

In general ρ_0 and v_0 will be functions of position, and equations (6-2) through (6-4) are not easily solved. The pioneering work by Jeans made a drastic simplification of the problem by assuming that ρ_0 and C are constants and that v_0 vanishes. On this assumption equations (6-2) and (6-3) are simplified; if one takes the divergence of equation (6-2), and uses the resultant equation to eliminate $\mathbf{V} \cdot \mathbf{v}_1$ from equation (6-3), eliminating $\nabla^2 \phi_1$ by use of equation (6-4), the resultant equation for ρ_1 may be solved very simply. For an assumed plane wave of the form

$$\rho_1 = K e^{i(\kappa x + \omega t)} \tag{6-5}$$

one gets

$$\omega^2 = \kappa^2 C^2 - 4\pi G \rho_0 \tag{6-6}$$

If the wave number κ is less than a critical value, κ_J, ω^2 is negative, and the disturbance grows exponentially. Hence any condensation of length greater than π/κ_J will be gravitationally unstable. This critical length, L_J, called the "Jeans length," is given by

$$\kappa_J^2 = \left(\frac{\pi}{L_J}\right)^2 = \frac{4\pi G \mu \rho_0}{kT} \tag{6-7}$$

While this approach has the advantage of simplicity, it does not represent a solution of any physical problem, since the assumed equilibrium conditions do not satisfy the basic equations. Equation (5-1) shows that $\nabla^2 \phi_0$ must vanish in a hypothetical uniform homogeneous medium with v_0 zero. However, equation (5-3) indicates that $\nabla^2 \phi_0$ cannot vanish if ρ_0 is finite. While equation (6-7) can be justified by dimensional analysis, the numerical constant cannot be trusted; in some cases, the results obtained with the Jeans method can be misleading qualitatively as well as quantitatively [2].

6. FORMATION OF STARS

Instability of a Gaseous Disc

The simplest problem that permits an exact analysis and shows gravitational instability is the perturbation of a uniform gaseous disc of infinite radius, subject to its self-gravitational attraction. In the equilibrium state $\mathbf{v}_0$ is assumed to vanish, while ρ_0 and ϕ_0 are functions of z only. Equations (5-1), (5-3), and (5-6) may be combined to give

$$\frac{d}{dz}\left(\frac{1}{\rho_0}\frac{d\rho_0}{dz}\right) = -\frac{4\pi G}{C^2}\rho_0 \qquad (6\text{-}8)$$

The solution of this equation is [3]

$$\rho_0 = \rho_0(0)\operatorname{sech}^2 \frac{z}{H} = \rho_0(0)(1 - w^2) \qquad (6\text{-}9)$$

where $\rho_0(0)$ is the value of the unperturbed density, ρ_0, at the midplane, $z = 0$;

$$H \equiv \frac{M}{2\rho_0(0)} = \left(\frac{kT}{2\pi G\mu\rho_0(0)}\right)^{1/2} \qquad (6\text{-}10)$$

and

$$w \equiv \tanh\frac{z}{H} = \frac{M(z)}{M} \qquad (6\text{-}11)$$

For positive z, the quantity $M(z)$ is the mass of the disc per unit area between $-z$ and $+z$, while M denotes $M(\infty)$, the total mass of the disc per unit area.

Instead of solving the linearized equations, we consider now the perturbation which neither oscillates nor grows exponentially, corresponding to ω equal to zero in equation (6-5). Such a state of "marginal stability" generally separates regions of stability, where small perturbations are damped, from instability regions, where small perturbations grow exponentially. Situations in which instability occurs by exponential growth of oscillations (called "overstability") must be treated by other means [4]. In general overstability does not appear in conservative systems (in which viscosity, resistivity, and thermal conductivity vanish), provided that $\mathbf{v}_0$, the velocity in the unperturbed state, is zero or can be transformed away. Since we are considering here a conservative system, with $\mathbf{v}_0$ equal to zero, the marginal state in which the perturbed quantities are independent of time will separate regions of stability and instability.

Defining κ_c as the value of κ in the marginal state, we may write for this state

$$\frac{\rho_1(w)}{\rho_0(w)} = e^{i\kappa_c x}\theta(w) \tag{6-12}$$

independent of t and of the coordinate y; the variable w defined in equation (6-11) replaces z. In equation (6-2) the left-hand side vanishes in the marginal state; if the divergence of this equation is taken, with $\nabla^2 \phi_1$ eliminated by use of equation (6-4), one obtains [5] after straightforward calculation, using equations (6-9) and (6-11),

$$\frac{d^2\theta}{dw^2} - \frac{2w}{1-w^2}\frac{d\theta}{dw} + \theta\left\{\frac{2}{1-w^2} - \frac{v^2}{(1-w^2)^2}\right\} = 0 \tag{6-13}$$

where

$$v = \kappa_c H \tag{6-14}$$

The general solution of equation (6-13) is

$$\theta(w) = A_1\left(\frac{1+w}{1-w}\right)^{v/2}(v-w) + A_2\left(\frac{1-w}{1+w}\right)^{v/2}(v+w) \tag{6-15}$$

Since $\theta(w)$ must remain finite as w approaches ± 1, v must equal unity. From equations (6-14) and (6-10) we obtain

$$\kappa_c^2 = \frac{1}{H^2} = \frac{2\pi G \mu \rho_0(0)}{kT} \tag{6-16}$$

Physically one would expect instabilities to appear on the long-wavelength side of the marginal state, and we may conclude that gravitational instability appears for κ less than κ_c. The minimum length of an unstable region in the plane of the disc is about πH; since the thickness of the disc is somewhat less than this value,

6. FORMATION OF STARS

motions in the z direction only cannot produce instability, as may be shown directly from the linearized equations [5].

In the classical Jeans analysis the amplification rate, $-i\omega$, increases to a constant value, $(4\pi G \rho_0)^{1/2}$, as κ falls below κ_J [see equation (6-6)]. In the present case, however, $-i\omega$ rises to a maximum value for κ about half κ_c [6] and falls toward zero as κ approaches zero, corresponding to the physical fact that the potential gradient produced by density perturbations of constant amplitude in an infinite disc does not increase as the wavelength in this plane grows indefinitely.

When the disc has a finite radius and rotates around its axis the situation is more complicated. It is well known that the density of a rotating gas must generally exceed some lower limit to permit gravitational contraction. For example, the condition that the centrifugal force at the equator of a gaseous sphere of radius r and uniform density ρ rotating with the angular velocity, Ω, is less than the self-gravitational force gives immediately

$$\rho > \frac{3\Omega^2}{4\pi G} \tag{6-17}$$

The detailed analysis confirms that rotation tends to hinder the growth of condensations in a rotating disc [7, 8]. We have already seen that the gravitational condensations in a non rotating disc are somewhat flattened; this effect weakens self-gravitation as compared to the disruption force of rotation, and increases by a factor of about 4 the minimum density required for gravitational instability. When one considers also the shearing effect of galactic rotation, resulting from the variation of $\Omega(R)$ with R, the following condition for instability is found [8] for an isothermal disc,

$$\rho_0(0) > \frac{4.4\Omega^2}{\pi G}\left(1 + \frac{R}{2\Omega}\frac{d\Omega}{dR}\right) \tag{6-18}$$

With somewhat different assumptions the constant 4.4 in condition (6-18) is increased [7] to about 6. As $\rho_0(0)$ increases above the critical value in condition (6-18), instability first sets in at a wavelength about twice $2\pi H$. At this wavelength the e-folding rate, $-i\omega$, is a maximum in the absence of rotation; the band of unstable wavelengths broadens as $\rho_0(0)$ increases further. Condition (6-18) refers to ring-shaped condensations, in which all motions are radial. Condensations in which the motions are tangential will be distorted by the shearing effect of galactic rotation, but for a short period these "sheared condensations" can apparently grow exponentially [8] even when the gas is stable against the radial condensations.

If these conditions are applied to the Galaxy in the neighborhood of the Sun, where Ω is about 8×10^{-16} sec^{-1}, and $R\,d\Omega/2\Omega\,dR$ is about -0.6 [9], condition (6-18) yields as a condition for instability

$$\rho_0(0) > 5.4 \times 10^{-24} \text{ g/cm}^3 \qquad (6\text{-}19)$$

nearly three times the values adopted in Table 5.1 for the mean interstellar density. As we have seen in Section 1.2, a gas density between 5 and 10×10^{-24} g/cm^3 cannot be excluded, and, as noted above, sheared condensations can grow for a while even if $\rho_0(0)$ is below the critical value. Moreover, the presence of stars, with a mean density of at least 4×10^{-24} g/cm^3 (see Section 1.2) may permit gravitational instability even if the gas density does not satisfy inequality (6-19), although a much more complex analysis is needed to analyze the role of stars in this respect.

In any case the condensations produced by such instabilities would be very large, with diameters of at least 300 pc, and masses of some $10^6 M_\odot$. As we shall see below, magnetic forces may modify the formation of these large condensations, though it seems likely that instabilities can develop parallel to the magnetic field, provided that **B** is parallel to the galactic plane. These large condensations may play an important role in the formation of spiral arms, apparently the first step in star formation. Alternatively, the spiral arms may represent instabilities in the distribution of stars, with the gas playing a secondary role.

Collapse of an Isolated Cloud

The previous discussion concerns a gravitationally bound system in equilibrium, which is unstable against the formation of several

6. FORMATION OF STARS

smaller condensations. Here we discuss a different type of condensation, in which a system is in equilibrium under certain conditions, but as conditions change equilibrium is no longer possible and the system contracts as a unit. Since the zero-order equations have no steady-state solution in this case, the contraction is not really a true instability at all, and we shall refer to it as a collapse.

The simplest interstellar situation in which gravitational collapse appears is an isothermal nonrotating sphere embedded in a medium of constant pressure, p_0. An intercloud medium at a temperature much higher than within the cloud, and at a correspondingly lower density, will be a close approximation to such an idealized constant-pressure medium. The equilibrium condition for such a sphere may be obtained from the virial theorem, equation (5-11). We ignore at first the magnetic field, but the pressure term at the surface of the sphere, at radius R from the center, must be retained. In a steady state the velocity and the acceleration both vanish. Since the mass outside the sphere may be ignored, W equals the gravitational energy, and if we approximate this quantity by the usual expression for a sphere of uniform density, we have

$$4\pi R^3 p_0 = \frac{3MkT}{\mu} - \frac{3GM^2}{5R} \qquad (6\text{-}20)$$

When p_0 is very small, this equation permits a stable solution for which R is large and the gravitational binding is negligible. As p_0 increases, with M kept constant, the sphere contracts, the density increases, and the gravitational binding becomes more important, decreasing the radius somewhat further. As R varies, with M constant, equation (6-20) gives an upper limit for p_0, which we designate as p_m. Straightforward computations give

$$p_m = 3.15 \left(\frac{kT}{\mu}\right)^4 \frac{1}{G^3 M^2} \qquad (6\text{-}21)$$

Use of the gravitational potential for a uniform sphere underestimates W, since the sphere is somewhat condensed toward the center when p_0 approaches p_m. It is readily seen that in the actual case [10] p_m will be somewhat less than given by equation (6-21). A detailed solution [1] for the structure of the isothermal sphere with constant pressure at the boundary yields a numerical constant 1.40 instead of 3.15 in equation (6-21). If p_0 exceeds p_m, no solution is possible, and the sphere will collapse.

This type of isothermal collapse is possible only in three-dimensional systems, where the gravitational acceleration, g, varies more rapidly with R than does the pressure gradient. For the contraction of an isolated plane sheet, in a direction perpendicular to the plane of the sheet, g for any layer is unaffected by changes in the sheet thickness, while $\nabla p/\rho$ increases as the sheet is compressed, provided the gas temperature is constant. Thus the sheet is stable for such perturbations, regardless of the value of the external pressure. A change of temperature with density can produce a thermal instability in this case, if T decreases more rapidly than $1/\rho$ as ρ increases. The two-dimensional case is intermediate in that the gravitational force, $\rho \mathbf{g}$, and the pressure gradient, ∇p, change proportionally in the course of isothermal, cylindrical contraction. Collapse of an isothermal cylinder cannot be produced by increasing the external pressure, but can result either by increasing the mass per unit length above a critical value [10] or by a slight decrease of T with increasing ρ; this latter contraction is really a thermal instability rather than a gravitational collapse.

The effect of a magnetic field may be included approximately. We assume that the field within the cloud is uniform, with a strength B_0, and that outside the cloud the mean square field strength for each value of r equals $B_0^2 (R/r)^6$. If the closed surface, S, in equation (5-11) is taken to be far outside the cloud, the magnetic surface terms are negligible, and the fields inside and outside the cloud contribute equally to $\mathscr{M}$. Calculations based on more realistic assumptions, with continuous field derivatives, show [11] that the internal field contributes somewhat more to $\mathscr{M}$, and the external field, somewhat less, the total $\mathscr{M}$ remaining about the same. The sum of the other terms is the same as when the surface S coincides with the cloud surface, since the increase of U

6. FORMATION OF STARS

just offsets the increase of the surface integral involving p_0. Equation (5-11) then yields, in place of equation (6-20)

$$4\pi R^3 p_0 = 3\frac{MkT}{\mu} - \frac{1}{R}\left(\frac{3}{5}GM^2 - \frac{1}{3}R^4 B_0^2\right) \quad (6\text{-}22)$$

If the field is assumed to be frozen in the material, the total flux through the cloud, equal to $\pi R^2 B_0$, must remain constant as the cloud contracts; consequently the gravitational and magnetic energies change proportionally as the cloud is compressed. If magnetic forces do not prevent initial contraction, they will not prevent a spherical collapse at any stage. This same conclusion also follows from comparing the magnetic force perpendicular to **B** and the gravitational force; if expressed per cubic centimeter, both forces vary as $1/R^5$.

The quantity R^4 in parentheses in equation (6-22) may be expressed in terms of M, with ρ taken to be constant. It then follows that the gravitational term exceeds the magnetic term and that collapse is possible only if

$$M > M_c \equiv \frac{5^{3/2}}{48\pi^2}\frac{B^3}{G^{3/2}\rho^2} = 0.0236\frac{B^3}{G^{3/2}\rho^2} \quad (6\text{-}23)$$

and the maximum pressure, p_m, in equilibrium, and the corresponding radius, R_m, are given by

$$p_m = 1.40\frac{(kT/\mu)^4}{G^3 M^2\{1-(M_c/M)^{2/3}\}^3} \quad (6\text{-}24)$$

$$\frac{GM^2}{R^4} = \frac{25 p_m}{1-(M_c/M)^{2/3}} \quad (6\text{-}25)$$

In equations (6-24) and (6-25) the numerical constants have been taken from the exact solution [1] for no magnetic field, which may give a better approximation for finite B than the approximate constants obtained directly from equation (6-22).

These numerical results may be applied to conditions in the Galaxy. For the density ρ we shall take 4×10^{-23} g/cm², twice the value in Table 5.1, based on the standard cloud parameters in Table 3.4, but a not unreasonable value for a large cloud (see

Table 3.2). If B equals 3×10^{-6} G, M_c is then $1.2 \times 10^4\ M_\odot$. These values suggest that clouds of the observed density are more likely to condense into clusters or associations than into isolated stars. On the other hand, for a density equal to 6×10^{-24} g/cm^3, about equal to the minimum density for instability of the galactic disc, according to equation (6-19), then M_c rises to $5 \times 10^5\ M_\odot$ for the same assumed field. Since this value of M_c is comparable with the masses required for gravitational instability in the galactic disc, magnetic forces may affect the large-scale galactic instabilities discussed above.

Examination of the types of clouds listed in Table 3.2 shows that the large cloud may have a mass exceeding somewhat the critical mass above. If these clouds are on the verge of collapse, then M, R, and p_0 should be related by equation (6-25) with p_0 assumed to equal p_m, its maximum value. If p_0 is set equal to 8×10^{-14} dynes/cm^2, corresponding to the cloud parameters in Table 5.1, or to a density of 4×10^{-23} g/cm^2 and a temperature of $30°$, the theoretical value of $M/\pi R^2$, the mass per unit area, is 2.9×10^{-3} g/cm^2; M/M_c has been assumed to equal 2. The corresponding mean extinction would be about 1 magnitude, roughly that observed. However, the mean extinction in two of the other three clouds listed in Table 3.2 is also of about this order, though the masses are much smaller than the value of M_c found above. It should be noted that if B were as great as 2×10^{-5} G, M_c would be increased to $4 \times 10^6\ M_\odot$, for ρ again equal to 4×10^{-23} g/cm^3, and there would seem little prospect of gravitational collapse for any of the observed types of clouds [1].

Evidently one can infer that present knowledge is not inconsistent with the belief that the larger interstellar clouds are near the state where gravitational collapse can occur. Possibly an increase in their mass by accretion, an increase of the external pressure by a passing shock wave, or a decrease in their internal turbulence could start the collapse process. On the other hand. new information on the magnetic topography or on other properties of the interstellar medium could well lead to an entirely different picture of the beginning stage in star formation.

6. FORMATION OF STARS

6.2 Fragmentation of a Collapsing Cloud

The previous section has treated the conditions under which interstellar clouds might begin to contract under gravitational forces. In the present section the accelerating collapse is considered. The chief emphasis here is on the way in which a large gas cloud might be expected to break up into individual condensations, a process called fragmentation. We have already seen that gravitational condensation does not seem possible for clouds of solar mass, unless the magnetic field in the clouds is very much less than is generally supposed. Hence fragmentation of a more massive condensation into structures of stellar mass must probably be assumed at some stage of star formation. This section analyzes the possibility that fragmentation begins during the initial gravitational collapse.

In this complex problem it is not possible to make any very definite statements about what actually happens in interstellar space. Again the discussion is devoted to idealized models. First we consider the collapse of a nonrotating cloud with no magnetic field, treating subsequently the effects produced either by rotation or by a magnetic field. Some discussion of the effects that may be expected when both angular momentum and magnetic flux are present appears in the final section of this chapter.

Uniform Nonrotating Cloud, $\mathbf{B} = 0$

A spherical gas cloud of radius R is assumed to start collapsing, its temperature remaining constant. As the radius decreases, the gravitational force per cubic centimeter will go up as $1/R^5$, while the isothermal pressure gradient varies as $1/R^4$. Hence after a moderate decrease in R the pressure is no longer important, and the material accelerates as though it were almost completely cold, and falls freely towards the center. The rise in temperature which may result when the opacity has increased sufficiently is considered in Section 6.3. We compute here the time of collapse for a cold sphere on the assumption that the density, $\rho(t)$, is independent of radius, and that the gas is initially motionless. Since the analysis is exact, the subscripts 0 and 1 introduced in the preceding section are omitted in the following paragraph.

We consider a particular mass shell, and let $r(t)$ be its radius as a function of time. If a denotes, for this particular mass shell, the value of r initially, when free-fall is assumed to begin, the equation of motion is

$$\frac{d^2 r}{dt^2} = -\frac{GM(a)}{r^2} = -\frac{4\pi G \rho(0) a^3}{3 r^2} \tag{6-26}$$

where $M(a)$ is the mass interior to the initial radius; evidently the mass inside the shell stays constant during the collapse, if the shells are assumed not to cross each other. The energy integral of equation (6-26) gives

$$\frac{dr}{a dt} = -\left\{\frac{8\pi G \rho(0)}{3}\left(\frac{a}{r}-1\right)\right\}^{1/2} \tag{6-27}$$

If we make the substitution

$$r/a = \cos^2 \beta, \tag{6-28}$$

equation (6-27) yields [12]

$$\beta + \tfrac{1}{2}\sin 2\beta = t\left\{\frac{8\pi G \rho(0)}{3}\right\}^{1/2} \tag{6-29}$$

where we let t vanish initially, when dr/dt is zero. Evidently β is the same for all mass shells at any one time, and all shells reach the center at the same time, when β equals $\pi/2$, and after a "free-fall time" given by

$$t_f = \left\{\frac{3\pi}{32 G \rho(0)}\right\}^{1/2} \tag{6-30}$$

If the initial density is 4×10^{-23} g/cm^3, t_f is 1.1×10^7 years.

During this free fall the gas will tend to be unstable against the formation of smaller condensations. If we recall that the temperature is actually finite, not zero, the size of a condensation that is just unstable may be estimated from the usual Jeans length.

6. FORMATION OF STARS

We now denote by $\rho_0(t)$ the density in the uniformly contracting sphere, with $\rho_1(t)$ representing the excess density in the smaller condensations. The minimum mass of the condensation that is just unstable is proportional to $\rho_0 L_J{}^3$, and according to equation (6-7) varies as $1/\rho_0^{1/2}$. When the collapse starts, the sphere as a whole will contract, but as ρ_0 increases, progressively smaller masses may be expected to become gravitationally unstable [13]. For example, when the radius is one-tenth its initial value, the density has increased by a factor 10^3, and the mass of the condensation which is just unstable has decreased by a factor 30. To evaluate the importance of this fragmentation process requires quantitative information on the growth of condensations of different sizes in a contracting sphere.

While this problem has not been fully analyzed, the growth of infinitesimal instabilities in a completely cold gas has been treated [12]. This analysis does not take into account the breakup of the cloud into smaller and smaller units as the collapse proceeds but does give the rate of growth in the most favorable case where the pressure is negligible and as a result disturbances of all wavelengths are equally unstable.

This analysis starts with the contracting cold gas sphere analyzed above as the zero-order solution and examines small displacements from this solution exactly as in Section 6.1. From equation (6-28) we see that the zero-order velocity and density are given by

$$\mathbf{v}_0 = -2\mathbf{r} \tan \beta \frac{d\beta}{dt} = -\mathbf{r} \sin \beta \sec^3 \beta \left\{ \frac{8\pi G \rho(0)}{3} \right\}^{1/2} \quad (6\text{-}31)$$

$$\rho_0(t) = \rho_0(0) \sec^6 \beta \quad (6\text{-}32)$$

where equation (6-29) has been used to evaluate $d\beta/dt$. If we take the divergence of the linearized momentum equation (6-2), with the sound velocity, C, set equal to zero, we obtain

$$\frac{\partial}{\partial t}(\nabla \cdot \mathbf{v}_1) + \mathbf{v}_0 \cdot \nabla(\nabla \cdot \mathbf{v}_1) - 4 \tan \beta \frac{d\beta}{dt}(\nabla \cdot \mathbf{v}_1) = -\nabla^2 \phi_1 \quad (6\text{-}33)$$

where we have used equation (6-31) and the vector identities

$$\mathbf{V} \cdot (\mathbf{r} \cdot \mathbf{V}\mathbf{v}_1) = \mathbf{r} \cdot \mathbf{V}(\mathbf{V} \cdot \mathbf{v}_1) + (\mathbf{V} \cdot \mathbf{v}_1) \qquad (6\text{-}34)$$

$$\mathbf{v}_1 \cdot \mathbf{V}\mathbf{r} = \mathbf{v}_1 \qquad (6\text{-}35)$$

Since ρ_0 is independent of $\mathbf{r}$, and $\mathbf{V} \cdot \mathbf{r}$ equals 3, the linearized equation of continuity, (6-3), becomes.

$$\frac{\partial \rho_1}{\partial t} + \mathbf{v}_0 \cdot \mathbf{V}\rho_1 = 6\rho_1 \tan \beta \frac{d\beta}{dt} - \rho_0 (\mathbf{V} \cdot \mathbf{v}_1) \qquad (6\text{-}36)$$

Equations (6-4), (6-33), and (6-36) determine the perturbed motion.

To solve these three equations we make the substitution

$$\mathbf{V} \cdot \mathbf{v}_1 = H(\beta) \sec^4\beta \, g(\mathbf{a}) \qquad (6\text{-}37)$$

where instead of t and the Eulerian coordinate, $\mathbf{r}$, we have used as independent variables β, a function of time through equation (6-29), and the Lagrangian coordinate, $\mathbf{a}$, defined as the position vector of a fluid element initially, when β and t vanish; the scalar value of $\mathbf{a}$ is the initial radius of a mass shell, as defined above. The separation of time and space variables in equation (6-37) implies consideration of standing waves. The amplitude function, $g(\mathbf{a})$, is arbitrary, except for the condition that $g(\mathbf{R})$ must vanish to avoid deformation of the outer surface [14], which is ignored here. In this Lagrangian coordinate system the terms in $\mathbf{v}_0 \cdot \mathbf{V}(\mathbf{V} \cdot \mathbf{v}_1)$ and $\mathbf{v}_0 \cdot \mathbf{V}\rho_1$ in equations (6-33) and (6-36) go out. Equation (6-4) now yields

$$\rho_1 = \frac{\nabla^2 \phi_1}{4\pi G} = -\left\{\frac{\rho(0)}{24\pi G}\right\}^{1/2} \sec^6 \beta \frac{dH}{d\beta} g(\mathbf{a}) \qquad (6\text{-}38)$$

where we have used equation (6-33), together with equation (6-37), to eliminate $\nabla^2 \phi_1$; in evaluating the time derivative in equation (6-33), the value of $d\beta/dt$ has been substituted from equation (6-29). If now equation (6-38) is used to eliminate $\partial \rho_1/\partial t$ from equation (6-36), evaluating the time derivative as before, and

6. FORMATION OF STARS

substituting from equation (6-32) for ρ_0, one gets

$$\frac{d^2H}{d\beta^2} = 6H \sec^2 \beta \tag{6-39}$$

The solution which increases with time is

$$H(\beta) = A(3 \sec^2 \beta - 2) \tag{6-40}$$

where A is an arbitrary constant. Equation (6-38) now yields

$$\frac{\rho_1}{\rho_0} \propto \sec^3 \beta \sin \beta \, g(\mathbf{a}) \tag{6-41}$$

This result indicates that the perturbed density grows more rapidly than $\rho_0(t)$, their ratio varying as $\rho_0^{1/2}$ as β approaches $\pi/2$. If the initial disturbances are about 10% of the mean density, then to make ρ_1 comparable with ρ_0, ρ_0 must increase by a factor of 100. Consideration of the finite pressure, which reduces the growth rate when the condensation first becomes unstable, results [12] in about a tenfold further increase in ρ_0 required under these same assumptions. We have already seen that the minimum mass which is just unstable varies as $\rho_0^{-1/2}$. Hence fragmentation into 30 subunits for each thousandfold increase in mean density may just about be possible if the initial small-scale irregularities present in the gas exceed about 10% of the mean density.

The foregoing analysis is concerned only with the formation of subcondensations. Clearly there will be a tendency for these individual subcondensations to collide with each other and to coalesce as the entire cloud contracts. Because of initial turbulence, different regions of the cloud will not pass exactly through the center, and because of the nonuniform density distribution different parts of the cloud will experience their closest approach to the center at different times. Nevertheless, the density in the central zone will be relatively high, and what fraction of the subcondensations can survive once they have formed is controversial [14, 15]. In general, to follow theoretically the course of fragmentation beyond its earliest beginnings is very difficult, and all such discussions are subject to much uncertainty.

Rotating Cloud, **B** $= 0$

If an isolated cloud has initially some angular momentum, **J**, this will remain constant during the contraction. If $\Omega(t)$ is the angular velocity at the equator, and the cloud is assumed to contract homologously, then the constancy of **J** yields

$$\Omega(t)\,R^2(t) = \Omega(0)\,R^2(0) \qquad (6\text{-}42)$$

where $R(t)$ is the cloud radius. The centrifugal force at the equator per unit mass equals $\Omega^2 R$, which in accordance with equation (6-42) varies as $1/R^3$, while the gravitational force per gram varies only as $1/R^2$. At a certain equilibrium radius, R_e, centrifugal and gravitational forces are equal at the equator. The cloud cannot contract much below R_e without leaving some mass at about this radius.

Centrifugal force operates, of course, only in the plane perpendicular to **J**. Parallel to **J** contraction can proceed as before, except that the accelerating force no longer increases as the system contracts in one dimension. This contraction parallel to **J** will not much affect the radius of the disc perpendicular to **J**, since the gravitational force in this direction is not greatly increased by contraction of the sphere to a disc. Depending on the extent to which random motions are damped out, the cloud can become a highly flattened disc or an oblate spheroid [2].

While angular momentum can stop the collapse of the cloud in two dimensions, it need not, in principle, terminate the process of fragmentation. If a gaseous disc is assumed to form, the density in the disc will increase with continuing contraction. We have seen in Section 6.1 that when the gas density in the plane of the disc, at a distance R from the axis, exceeds a critical value, the gas in the disc becomes gravitationally unstable. Subcondensations can form within the disc, and these can continue the process of contraction and fragmentation. Depending on the relative rates

6. FORMATION OF STARS

of velocity dissipation, fragmentation, and coalescence, this process might end with all fragments flattened and rotating in the same plane or, at the other extreme, with the fragments each moderately oblate and forming a system that was only slightly flattened.

While fragmentation can continue, in principle, until small masses are formed, it does not appear that stars can be formed in this manner. Since each successive fragment contracts primarily in a direction parallel to **J**, with less contraction in the transverse direction, this process is effectively a method for a one-dimensional collapse parallel to **J**, triggered by an initial condensation perpendicular to **J**. To consider an extreme case, let us assume that the material for each subcondensation comes from a cylinder which extends across the initial cloud, whose axis is parallel to **J**, and which contracts to form a sphere of stellar density. This assumption yields condensations with the minimum amount of angular momentum. However, the thickness of the cloud is limited, and to provide enough mass for a star a great deal of transverse contraction, perpendicular to **J**, is also needed. For example, in the large cloud in Table 3.2 a cylinder 0.1 pc in radius extending 40 pc in a direction parallel to the rotation axis contains about one solar mass. However, to form a star of solar density the radius must decrease by 2×10^{-7}, giving an increase of Ω by 3×10^{13} from equation (6-42). If the original angular velocity of the large cloud is assumed to be about the same as that for the Galaxy, a not unreasonable assumption as regards order of magnitude, the period of rotation for the star formed in this way would be roughly 5 min, and centrifugal force would exceed gravity by some 3 orders of magnitude. Under these conditions objects of 10^{-5} $M_\odot$ could be gravitationally bound [2], but not a normal star. The angular momentum within a protostar, relative to its own center of gravity, must apparently be reduced by almost 2 orders of magnitude to produce stars of the type observed. In principle, the angular momentum can either be concentrated in relative motion between different stellar masses, as in a multiple star system, or be transferred outside the stellar system entirely.

Uniformly Magnetized Cloud, $\Omega = 0$

A uniform magnetic field within a cloud has a certain similarity to rotation in that both B and Ω vary as $1/R^2$ for contraction of the cloud in a direction perpendicular to **B** or to Ω. The energy densities per cm^3 are entirely different in the two cases, however, varying as $1/R^4$ in the former, and $1/R^5$ in the latter, assuming isotropic contraction. We have already seen that as a result of this difference these two fields have somewhat dissimilar effects on at least the initial phases of gravitational collapse; the magnetic field leads to a lower limit on the mass which can collapse, while the rotation gives rise to a lower limit on the density of a cloud which can start to collapse. It will be shown here that with respect to continuing fragmentation, on the other hand, the two fields are rather similar, in that contraction into discs perpendicular to **B** permits fragmentation to continue in somewhat the same fashion as discussed above for the rotating cloud.

One may consider the collapse of a cold magnetized cloud, whose mass, M, is assumed somewhat greater than the critical mass, M_c, defined in equation (6-23). The ratio of the magnetic and gravitational energies in the cloud equals $(M_c/M)^{2/3}$, as may be verified from the ratios of the last two terms in equation (6-22). The magnetic energy opposes the acceleration perpendicular to **B**; thus if M/M_c equals 4, for example, $(M_c/M)^{2/3}$ is 0.4, and acceleration perpendicular to **B** may be reduced by some 40% relative to acceleration parallel to **B**. Evidently the sphere will tend to flatten, forming an oblate spheroid, at the same time as it contracts. This flattening tendency will be heightened by the somewhat increased gravitational force at the poles of a flattened spheroid with respect to the equator [2].

This flattening complicates the dynamical analysis, since both the gravitational and magnetic energies will change by factors of 2 to 3 as the system flattens [11]. To demonstrate the possible continuation of fragmentation it suffices to show that M_c decreases in this process. For spherical contraction M_c remains constant, since B varies as $\rho^{2/3}$ when the gas is compressed isotropically.

6. FORMATION OF STARS

However, when a cloud contracts in a direction parallel to **B**, ρ increases without change of B, and M_c decreases.

This change of M_c can also be indicated by an alternative form of equation (6-23). If we replace M by $4\pi R^3 \rho/3$, assuming a uniform sphere, we obtain as a necessary condition for gravitational collapse of a sphere

$$R\rho > \frac{5^{1/2}}{4\pi} \frac{B}{G^{1/2}} \tag{6-43}$$

This inequality may be applied to the idealized case of an initially spherical cloud of radius R_1 and density ρ_1 that flattens to a disc of half-thickness h without contraction across the field. During this flattening B remains unchanged, but the density increases to a value ρ_2, where

$$\rho_2/\rho_1 = R_1/h \tag{6-44}$$

Let us consider whether a spherical condensation of radius R_2 can form in the disc. If by analogy with the results on gravitational instability in Section 6.1 we assume that the radius of the condensation is about equal to h, then equation (6-44) yields

$$R_2 \rho_2 = R_1 \rho_1 \tag{6-45}$$

It follows from equation (6-43) that if the initial cloud is collapsing, then the less massive fragments in the disc can also collapse gravitationally [2]. It also follows that if the mass of the spherical cloud is less than M_c, and the cloud is not collapsing gravitationally, then contraction parallel to the magnetic field will not make condensation of smaller units possible.

These results are subject to the general uncertainty which, as we have noted above, characterizes all discussion of continuing fragmentation. In particular, for a collapsing cloud it is probably not realistic to apply the criteria which were derived for equilibrium configurations. In the later stages of fragmentation, when the kinetic energy of implosion has become large, one would expect the shape of the subcondensations to depend more on the velocity field established earlier than on the instantaneous magnetic and gravitational forces. In addition, the topography of the magnetic field and the resultant magnetic forces may be much more complex than is taken into account in this simple analysis. One can conclude only that theory does not reveal any fundamental and decisive argument against the continuing fragmentation of a nonrotating magnetized cloud down to condensations of stellar mass.

As in a rotating cloud, the fragmentation of a magnetized cloud by means of preferential axial contraction effectively brings together into one object all the mass lying within some tube of force extending through the initial cloud. As we have seen before, a cylinder 40 pc long and 0.1 pc in radius in a cloud of density 4×10^{-23} g/cm^3, contains a total mass of about 1 $M_\odot$. The magnetic field in the star formed from this material will exceed that in the original cloud by 3×10^{13}, the same factor found for Ω. If B were 3×10^{-6} G in the initial cloud, in the star of solar type the magnetic field would equal 10^8 G, giving a magnetic energy about one-fourth the negative gravitational energy. The discrepancy between objects formed in this way and the observed stars is not nearly so great for a magnetized cloud as was found above for a rotating cloud.

6.3 Later Stages in Star Formation

As a cloud of interstellar gas collapses under its own gravitational force, a number of additional physical effects probably become important. The opacity of the gas increases with increasing density, and the temperature may start to rise. The lines of magnetic force, together with ionized atoms and free electrons, may drift out of the gas, reducing the magnetic flux through each element of the gas. Finally the angular momentum may be redistributed within the system or possibly even carried away entirely either by some escaping matter or by magnetic torques; we have seen that the angular momentum per unit mass associated with the rotation of a protostar must definitely be reduced in

6. FORMATION OF STARS

some way if a normal star is to be formed. These processes are discussed very briefly in this section.

Quite apart from the detailed mechanisms involved, the later phase in star formation may be defined as the period in which the total energy, the angular momentum, and probably also the net magnetic flux of the protocluster are being reduced. All three of these quantities are constant in the simple dynamical models discussed earlier in this chapter. Yet both the total energy and the angular momentum of the gas must be very much reduced before this material can constitute normal stars, and some reduction in magnetic flux is probably also required. Thus one may visualize an interval of time, probably much exceeding the free-fall time defined in equation (6-30), during which these three quantities decrease, permitting continuing contraction. For example, if B and Ω are both negligible, one might expect the cloud to undergo the same type of "Kelvin-Helmholtz contraction" that presumably characterizes the early life of a newly born star. In this phase the gas contracts adiabatically, with the rate of change of total energy equal to the rate of radiation from the surface. As we shall see below, there is some question whether a cloud of stellar mass or greater can experience this type of contraction, and forces produced by rotation and by magnetic fields may play a vital part in any slow late phase.

The previous section has indicated that the dynamical evolution of a contracting protocluster becomes very difficult to follow shortly after gravitational collapse has begun. The later stages of star formation, where additional effects become important, are even more difficult to analyze in detail. The theoretical models discussed here are designed to show the nature of the relevant physical mechanisms, and are not to be considered as well-defined stages through which the fragments of a protocluster must pass on their way to forming stars.

Radiative Decrease of Energy

As a gas contracts isothermally, the total thermal energy remains constant. However, the work done in compressing the gas is radiated, and the total energy, E, is decreased by an amount ΔE. Since the work done against the gas pressure, p_G, equals $-p_G \Delta V$, where the volume change of a fluid element is ΔV, the work done per gram is $p_G \Delta \rho/\rho^2$. If equation (5-6) is used for p_G, and an integration is carried out over the entire cloud, we find

$$\Delta E = \frac{kT}{\mu} \int (\Delta \ln \rho) \, dM = \frac{MkT}{\mu} \overline{\Delta \ln \rho} \qquad (6\text{-}46)$$

where a bar denotes an average over the mass, M, of the cloud. Since $\ln \rho$ cannot change by a very large amount, ΔE is only a few times greater than the thermal energy MkT/μ; for an interstellar gas temperature of about 50°, ΔE is enormously less than the negative gravitational energy of the stars into which a protocluster presumably condenses. We may conclude that the radiative loss of energy during an isothermal collapse is negligibly small. Most of the energy released by the gravitational contraction goes into macroscopic kinetic energy and magnetic energy.

This conclusion will be altered if the compression raises the gas temperature, increasing significantly the thermal energy, which can subsequently be radiated. Thus shocks in the gas can provide an effective agent for radiating energy, since they will produce an abrupt increase of temperature followed by radiative cooling. Alternatively the internal energy will be increased if the effective rate of radiation from the gas is decreased so that the compression during the gravitational contraction is adiabatic. As in the case of a shock, radiation can then occur relatively slowly following the compression. Adiabatic compression could be produced if the cooling time, t_T, were much greater than the free-fall time, t_f, corresponding to very low values of the emissivity, j_v. However, Table 4.11 indicates that the values of t_T are relatively short at the high densities anticipated in a contracting cloud. As in Kelvin-Helmholtz contraction, adiabatic compression can also occur if τ, the optical depth down to the center of the cloud, is very great so that the flow of radiation out of the cloud is materially impeded.

6. FORMATION OF STARS

This type of adiabatic compression requires a large enough value of τ so that the contraction velocity, $-dR/dt$, resulting from the radiative loss of energy is substantially less than v_f, the free-fall velocity. For an approximate calculation of this lower limit on τ, we assume radiative equilibrium, and write for the total luminosity

$$L = 4\pi R^2 \times \frac{4\pi}{3}\frac{dJ}{d\tau} \approx \frac{4\pi R^2 c p_R(0)}{\tau} \qquad (6\text{-}47)$$

where J is the mean intensity of radiation, $p_R(0)$ is the radiation pressure at the center and c is the velocity of light. In this approximate relation we have assumed that $dJ/d\tau$ is constant down to the center of the cloud. The contraction velocity, $-dR/dt$, is obtained by setting L equal to the rate of decrease of the total energy. If we assume that the cloud is a polytrope of index n [16], and that the contraction is homologous, the condition that $-dR/dt$ is less than v_f gives

$$\tau > K_n \frac{c}{v_f} \frac{p_R(0)}{P(0)} \qquad (6\text{-}48)$$

where $P(0)$ is the total pressure at the center and K_n is a constant which depends on n and which varies between 5 and 185 as n increases from 0 to 3; to within an order of magnitude we may set K_n equal to 30.

The ratio $p_R(0)/P(0)$ can be estimated approximately from the theory of spherical gas masses in hydrostatic equilibrium [16], on the assumption that this ratio is constant throughout the mass. With a mean molecular weight of 1.27, corresponding to both hydrogen and helium neutral and n_{He}/n_H equal to 0.1, $p_R(0)/P(0)$ increases from about 0.01 for a cloud of $1 M_\odot$ to about 0.3 for one of $10 M_\odot$. For v_f equal to 3 km/sec, corresponding to a cloud of solar mass at a density of 10^{-14} g/cm^3, inequality (6-48) then yields

$$\tau > \begin{cases} 2 \times 10^4, & M_{\text{cloud}} = M_\odot \\ 10^6, & M_{\text{cloud}} = 10 M_\odot \end{cases} \qquad (6\text{-}49)$$

For small mass this critical value of τ varies about as M_{cloud}^2, and decreases more rapidly than the actual optical depth with decreasing values of M_{cloud}.

Values of τ and of the resultant radiation rates have been computed [17], taking into account absorption by atoms, molecules, and grains. The results indicate that condition (6-49) is satisfied only for clouds with a mass below $0.1 M_\odot$, and hence clouds of normal stellar mass cannot experience a quasistatic contraction of the Kelvin-Helmholtz type. If no major sources of opacity have been omitted, one may conclude that gas in clouds of a solar mass or more behaves dynamically like cold matter during times of the order of t_f.

Apparently it would seem that the reduction in the total energy must occur by some means other than through quasistatic contraction. In addition to radiation by shocks there are other mechanisms not involving radiation, including loss of energetic particles and transmission of energy by Alfvén waves.

Decrease in Magnetic Flux

The scale of interstellar magnetic fields is so large that dissipation of flux by simple Ohmic resistance is usually much too slow to be of importance during 10^{10} years [18] (see Section 5.1). There is another mechanism which can reduce the magnetic flux through a cloud, provided the gas is an H I region, composed primarily of neutral atoms. The electrons and positive ions in a gas are constrained to follow the lines of magnetic force, but the neutral atoms are not. Hence it is possible for the charged particles and the lines of force to drift together out of the gas, leaving the neutral atoms behind. Collisions between neutral atoms and charged particles will limit this drift velocity.

In laboratory experiments the requirement of electrical neutrality can lead to diffusion of ions and electrons out of a gas at the same rate. This process is called "ambipolar" diffusion. This same name is frequently given to a drift of interstellar ions and electrons through a gas of neutral atoms, though the detailed mechanism is

6. FORMATION OF STARS

entirely different, and magnetic rather than electrostatic forces are responsible for keeping the ions and electrons together.

We compute here the magnitude of w_D, the drift velocity of the ions and the magnetic field relative to the gas of neutral atoms. The magnetic field provides the primary driving force for w_D, since these forces act only on the ions, and not directly on the neutral atoms. If the lines of force are assumed straight and parallel, this force equals $\nabla B^2/8\pi$ per cm^3, and in a quasisteady state must be balanced by the exchange of momentum in collisions, chiefly those between positive ions and neutral hydrogen atoms. This mean momentum exchange may be computed from the slowing-down time, t_s, given in equation (4-9). Equating these two forces per cm^3 yields

$$\frac{1}{8\pi}\nabla B^2 = \frac{n_i m_i w_D}{t_s} = n_i n_H \sigma_s m_H \langle u_H \rangle w_D \qquad (6\text{-}50)$$

where $\langle u_H \rangle$ is the mean random velocity of hydrogen atoms relative to positive ions, and n_i and m_i are the particle density and mass of the positive ions, which we assume to have about the mass of carbon, though under some conditions hydrogen ions may predominate; the appropriate value of σ_s is given in Table 4.1. Collisions of helium atoms with ions also contribute, but since their random velocity is only half that of hydrogen and their cross section is also about half (see Section 4.1), they increase the right-hand side of equation (6-50) by about 10%, and we shall neglect such collisions.

The equation for w_D assumes a particularly simple form for an infinite cylinder supported against its self-gravitational force by the magnetic force on the ions. The ions and the magnetic field will then expand relative to the neutral gas at the velocity w_D, with the neutral gas slowly contracting. Instead of analyzing these motions in detail we may compute an approximate "diffusion time," t_D, defined as the ratio of the cylinder radius, r, to the relative velocity, w_D, between neutrals and ions. If we assume that the density, ρ, is uniform throughout the cylinder, then we obtain

$$t_D \equiv \frac{r}{w_D} = \frac{\sigma_s \langle u_H \rangle}{2\pi G m_H} \times \frac{n_i}{n_H} \times \frac{1}{(1 + 4n_{He}/n_H)^2} \qquad (6\text{-}51)$$

where we have assumed that n_i/n_H, the ratio of the positive ion density to the overall hydrogen density, is much less than unity. The contribution of helium atoms to the gas density has been taken into account; as before, we shall let n_{He}/n_H equal 0.1. For a temperature of 50° and an ion mass of 12 m_H, $\langle u_H \rangle$ is 1.1×10^5 cm sec^{-1} [see equation (4-24)], σ_s equals 2.1×10^{-14} cm^2, according to Table 4.1, and we find

$$t_D = 5.3 \times 10^{13} n_i/n_H \text{ years} \qquad (6\text{-}52)$$

While the assumptions here are somewhat idealized, equation (6-52) should give at least the order of magnitude of the time required for magnetic flux to diffuse out of an interstellar cloud. If n_i/n_H is 5×10^{-4}, a minimum value for a normal H I cloud, the diffusion time t_D is 2.7×10^{10} years, too long to be of interest for interstellar conditions.

Within a relatively dense cloud, n_i/n_H may fall to a very low value, since the ultraviolet radiation responsible for most of the ionization will be completely absorbed in the outer layers when the cloud density has increased by a few orders of magnitude above its initial level. In addition, energetic particles will be unable to penetrate a cloud at high density; even for the clouds described in Table 3.2 the range of a 2-MeV proton is about equal to the cloud diameter.

In the absence of any ionizing agents, the ion density will fall relatively rapidly. According to equation (4-106) protons will collide with grains during a mean time of about 4×10^7 years in a typical cloud at a density about 15 times the average value in interestellar space generally. When the density has increased by a factor 10^3, this time will fall to 4×10^4 years, and if an appreciable fraction of ions are neutralized on the grains, the ion density will be reduced by a factor $1/e$ in a time of this order. Dissociative recombination, in which an electron recombines with a molecular

6. FORMATION OF STARS

ion, which dissociates into neutral atoms, may also be important, since the cross section of this process can be many orders of magnitude greater than for normal radiative capture (see Section 4.1).

One may assume that at high densities the electron and ion densities fall to an equilibrium value determined by thermal ionization and by the more energetic particles in the cosmic radiation, which can still penetrate the compressed cloud. According to equation (2-14) thermal ionization of potassium, with an ionization potential of 4.34 V, will set in above 1000°K, maintaining n_e/n_H at about 10^{-7}, but at lower T we may ignore thermal ionization. Let us assume that ionization of H atoms by cosmic rays is balanced by collisions of ions with grains; equation (4-43) then becomes

$$n_H \zeta_f = n_i n_g \langle w_i \rangle \sigma_g \qquad (6\text{-}53)$$

Since the fractional ionization of hydrogen is very small, $n(\text{H I})$ has been replaced by n_H in this expression. The value of $n_g \sigma_g / n_H$, the cross section of the grains per H atom, is unaffected by the density in the cloud, and may be set equal to its value for the interstellar medium generally; in Section 3.2 the mean $n_g \sigma_g$ was found to equal 3.2×10^{-22} cm^{-1}, and the mean value of n_H has been set equal to 0.85 cm^{-3}, in accordance with Section 5.1. The ions will mostly be protons (or a mixture of H_2^+ and H_3^+ ions [19] if the hydrogen is mostly molecular), for which $\langle w_i \rangle$ equals about 10^5 cm/sec at 50°. With the value of ζ_f in equation (4-44), equation (6-53) gives

$$n_i = 0.18 \text{ cm}^{-3} \qquad (6\text{-}54)$$

The value of n_i may be reduced by a factor of about 1/2 by the charge on the grains, which increases the collision cross section [see equation (4-94)]. On the other hand, n_i could be increased if the intensity of cosmic rays of intermediate energy—between 30 and 300 MeV—were substantially greater than assumed. Equation (6-54) is valid only when n_H has increased to about 10^5 cm^{-3}, an increase by about 10^4 over its initial value, since at lower densities radiative recombination is dominant.

We see from equations (6-54) and (6-52) that when n_H has increased to about 10^7 cm^{-3}, t_D will be about 10^6 years. This time is much longer than the free-fall time at these densities, but is a not unreasonable duration for a slow contraction phase in which U, B, and J are generally decreased. For even greater densities—about 10^9 cm^{-3} for a cloud of mass 100 $M_\odot$—the mass per projected area of the cloud will exceed 80 g/cm^2, and the absorption even of relativistic particles by the outer layers of the condensation will further reduce n_i. Electrical charges will be left on some of the grains even if n_i falls to zero, and the current carried by such grains can maintain the magnetic field; however, the resultant time constant for B is relatively short [20]. We conclude that a marked reduction of the magnetic flux through a protostar seems entirely possible, provided that this material spends an appreciable time in a cold dense state.

Decrease in Angular Momentum

A simple calculation in Section 6.2 indicated that the angular momentum must be reduced by roughly 2 orders of magnitude to permit formation of stars of solar mass and radius. In the past, a number of mechanisms have been proposed [1] for reducing the angular momentum. In particular, rotational fission of a gas cloud seems an important method, in principle, for transferring angular momentum from rotation to orbital motion. This mechanism has not been analyzed in detail. Since the presence of a magnetic field in the interstellar gas seems very likely, retardation of rotation by a magnetic field also seems probable, and will be considered briefly here.

It is easy to demonstrate that magnetic forces are adequate to slow down the rotation, transmitting the angular momentum from a condensation out to the surrounding interstellar gas. In fact, if the magnetic lines of force through the cloud remain connected to the outlying material, it seems clear that the cloud cannot

6. FORMATION OF STARS

rotate indefinitely with a higher angular velocity than the surrounding gas, since differential rotation will continue to twist and bend the lines of force until the magnetic stresses retard the rotation of the inner regions. Thus uncertainty as to the magnetic topography is the chief problem in the rotational retardation of a contracting condensation.

The time scale for retardation by a magnetic field may be computed with a very simple model [21], in which the magnetic field **B** is initially constant, and a condensation or fragment, taken to be a region of relatively high density of radius R within this field, is rotating about an axis parallel to **B**. The initial angular velocity of the condensation is Ω, while that of the medium outside the condensation is zero. Thus initially there is a discontinuity in Ω along the line of force at the condensation radius. Evidently the field becomes twisted, and one may show that the kink in the lines of force travels outwards along the lines of force at the Alfvén velocity, V_A. This outgoing wave will accelerate the surrounding material up to the angular velocity Ω. The amount of material accelerated per second in the cylindrical shell of radius r and thickness dr is $2\pi\rho r V_A \, dr$; multiplication by Ωr^2 gives the angular momentum gained by the shell. If the moment of inertia of the condensation equals $2MR^2/5$, its value for a sphere of uniform density, we have

$$\frac{2MR^2}{5}\frac{d\Omega}{dt} = -\pi\rho V_A R^4 \Omega \qquad (6\text{-}55)$$

where the cloud radius is assumed constant with time and solid-body rotation is assumed. The right-hand side has been multiplied by 2 in this equation to account for wave propagation in both directions away from the condensation. The time constant, t_B, for slowing down the rotation by magnetic forces may then be taken as

$$t_B \equiv -\frac{\Omega}{d\Omega/dt} = \frac{4M}{5(\pi\rho)^{1/2}BR^2} \qquad (6\text{-}56)$$

where equation (5-10) has been used to eliminate V_A in equation (6-55).

The ratio M/BR^2 is essentially $(M/M_c)^{1/3} G^{1/2}$ [see equations (6-22) and (6-23)] and is unaffected by the contraction of a cloud across the field, or by fragmentation, if the initial mass and the initial flux are both divided equally among a number of fragments. From equation (6-30) it follows that if M/M_c does not differ much from unity, t_B is roughly equal to the free-fall time, t_f, at the density ρ. However, this density outside the condensation is likely to be substantially less than the density inside, giving a value for t_B appreciably greater than t_f for the condensation itself. Again, we must apparently assume that the condensations remain as extended interstellar clouds for a period substantially greater than the free-fall time.

It is not possible to indicate whether the reduction in angular momentum or in magnetic flux is likely to be the limiting factor which fixes the duration of this extended interval between initial gravitational collapse and final formation of stars. Quite probably, in any one fragment **B** and **Ω** will not be exactly parallel, and contraction along different directions will be subject to different limitations, with centrifugal force alone acting in one direction and magnetic force alone in the other. The magnetic retardation of the rotation will depend very much on whether the lines of force stay connected to the surrounding material or whether they become reconnected, as a result of very intense Ohmic dissipation or instabilities in localized regions [2]. In any case, if flux loss and angular momentum transport determine the rate of contraction, one would expect the resultant fragments to have appreciable magnetic fields and angular moments.

If galactic clusters are assumed to form from interstellar clouds, or protoclusters, the mean density of a protocluster at any stage is not likely [22] to exceed very much the presently observed mean cluster densities, which are comparable with the densities in a large cloud. Hence fragmentation must occur without a very large increase in the protocluster density. It follows that much of the fragmentation must occur before the magnetic flux through each fluid element has been appreciably reduced, since ambipolar diffusion requires a high local density. Probably much of the fragmentation will precede the slow contraction phase and the reduction of the angular momentum per unit mass.

It should be emphasized again that all these indications are very tentative. It is possible that fragmentation does not occur at all during the initial gravitational contraction of an interstellar cloud, and that normal stars are formed either by the explosion of a very massive supernova which has condensed from an interstellar cloud [23] or from the contraction of primordial objects which were formed at the same time as the Galaxy but which for some reason have deferred their appearance as stars until recently [24]. While marked progress has been made in recent years in understanding some of the physical processes that might be important in star formation, it is not yet possible to assemble these processes in a unitary and conclusive theory.

References

1. L. Spitzer, in *Stars and Stellar Systems*, Vol. 7, Univ. of Chicago Press, Chicago, 1968, p. 1.
2. L. Mestel, *Quart. J., Roy. Astron. Soc., London*, **6**, 161 and 265 (1965).
3. L. Spitzer, *Astrophys. J.*, **95**, 329 (1942).
4. S. Chandrasekhar, *Hydrodynamics and Hydromagnetic Stability*, Oxford University Press, London, 1961, Chapter 1.
5. P. Ledoux, *Ann. d'Astrophys.*, **14**, 438 (1951).
6. R. Simon, *Ann. d'Astrophys.*, **28**, 40 (1965).
7. V. F. Savranov, *Dokl. Akad. Nauk, USSR*, **130**, 53 (1960); *Ann. d'Astrophys.*, **23**, 979 (1960).
8. P. Goldreich and D. Lynden-Bell, *Monthly Notices Roy. Astron. Soc.*, **130**, 125 (1965).
9. M. Schmidt, in *Stars and Stellar Systems*, Vol. 5, Univ. of Chicago Press, Chicago, 1965, p. 513.
10. W. H. McCrea, *Monthly Notices Roy. Astron. Soc.*, **117**, 562 (1957).
11. P. Strittmatter, *Monthly Notices Roy. Astron. Soc.*, **132**, 359 (1966).
12. C. Hunter, *Astrophys. J.*, **136**, 594 (1962).
13. F. Hoyle, *Astrophys. J.*, **118**, 513 (1953).
14. C. Hunter, *Astrophys. J.*, **139**, 570 (1964).
15. D. Layzer, *Astrophys. J.*, **137**, 351 (1963).
16. A. S. Eddington, *Internal Constitution of the Stars*, Cambridge University Press, England, 1926, p. 117.
17. J. E. Gaustad, *Astrophys. J.*, **138**, 1050 (1963).

18. L. Woltjer, in *Stars and Stellar Systems*, Vol. 5, Univ. of Chicago Press, Chicago, 1965, p. 531.
19. L. B. Loeb, *Basic Processes of Gaseous Electronics*, University of California Press, Berkeley, Cal., 1960, p. 577.
20. L. Spitzer, in *Origin of the Solar System*, Academic Press, New York, 1963, p. 39.
21. R. Ebert, S. von Hoerner and S. Temesvary, *Die Entstehung von Sternen durch Kondensation Diffuser Materie*, Springer, Heidelberg, 1960, p. 311.
22. V. C. Reddish, *Vistas in Astronomy*, 7, 173 (1965).
23. J. F. Bird, *Rev. Mod. Phys.*, 36, 717 (1964).
24. D. Layzer, *Ann. Rev. Astron. Astrophys.*, 2, 341 (1964).

Symbols

a		Radius of grain.
		Radiation density constant $= 7.56 \times 10^{-15}\,\text{erg}\,\text{cm}^{-3}\,\text{deg}^{-4}$.
A	A_{kj}	Einstein probability coefficient for downwards spontaneous transitions.
	A_ν	Interstellar extinction at frequency ν in magnitudes.
	A_V	Interstellar extinction in the visible.
	A_r	Recapture constant for electron–proton collisions, equation (4-26).
b		Galactic latitude (b^{II}).
	b_j	Ratio of particle density to that in "equivalent thermodynamic system," equation (2-15).
B	$\mathbf{B}$	Magnetic field in gauss; $B_\perp$, $B_\parallel$, components of $\mathbf{B}$ perpendicular and parallel, respectively, to line of sight.
	$B_\nu(T)$	Planck function for intensity of radiation in thermodynamic equilibrium, equation (2-4).
	B_{jk}	Einstein probability coefficient for induced radiative transitions.
c		Velocity of light, 2.998×10^{10} cm sec^{-1}.
C		Isothermal speed of sound, equation (5-6); C_{I}, C_{II}, values of C in H I, H II regions.
		Optical depth in line center, equation (2-40).
	$C(u)$	Collision probability per particle per sec for relative velocity u.
e		Charge of proton, 4.803×10^{-10} esu.
		Base of Napierian logarithms.
		Subscript referring to electrons; i.e., n_e, T_e.
E		Energy in ergs.

247

	E_{rj}	Energy of atom in level j in stage of ionization r.
	E_m	Emission measure in pc cm^{-6}, equation (2-58).
	E_{B-V}	Color excess on the B-V system.
	$\bar{E}_2$	Mean kinetic energy per photoelectron.
f		Subscript denoting field particles.
	$f(w)$	Velocity distribution function.
	f_r	Partition function for atom in stage of ionization r, equation (2-11).
	f_e	Partition function for a free electron, equation (2-13).
F	F_c, F_i	Fraction of space in the galactic plane occupied by clouds, or by H II regions.
	$F(\phi)$	Phase function for scattering of light by grains through an angle ϕ.
	F_r	Force of radiation pressure on grains, equations (5-87) and (5-94).
$\mathscr{F}$		Flux of radiation.
g		Acceleration of gravity; g_z, acceleration perpendicular to galactic plane.
		Subscript referring to interstellar grains; i.e., n_g.
	g_{rj}	Statistical weight of level j, for atom in stage of ionization r.
	g_{ff}, g_{fn}	Gaunt correction factors for free–free transitions (equation (2-46)) and for free–bound transition to level n.
G		Gravitational constant, 6.67×10^{-8} cm^3 sec^{-2} g^{-1}.
		Subscript referring to interstellar gas; i.e., p_G.
h		Planck constant, 6.625×10^{-27} erg sec.
H		Effective half-thickness of galactic disc.
		Net heating rate of interstellar grains; H_r, H_a, components of H due to absorption of electromagnetic radiation and atomic collisions.
H		Subscript referring to neutral H atoms; i.e., n_H, m_H.
i		Subscript referring to positive ions; i.e., n_i, m_i.
I	I_ν	Specific intensity of radiation at frequency ν, equation (2-1).
j		Subscript usually referring to lower level in an atomic transition.
	j_ν	Emissivity of matter per cm^3 at frequency ν.
J		Particle flux of ionizing photons per cm^2 per sec.
k		Boltzmann constant, 1.380×10^{-16} erg deg^{-1}.
		Number of clouds in the line of sight per kpc.
		Subscript usually referring to upper level in an atomic transition.
K		Arbitrary constant.

SYMBOLS

l		Galactic longitude (l^{II}).
		Parameter characterizing Maxwellian velocity distribution, equation (4-3).
L		Path length.
		Luminosity; $L_\nu d\nu$, stellar luminosity within the frequency interval $d\nu$.
m		Particle mass, usually electron mass unless otherwise specified.
		Index of refraction, Chapter 3.
M		Total mass.
	M_c	Critical mass below which gravitational collapse and fragmentation are impossible in a magnetic field, equation (6-23).
$\mathcal{M}$		Total magnetic energy of system, equation (5-12).
n		Density of particles per unit volume; n_e, n_i, n_a, and n_g, values of n for electrons, ions, neutral atoms, and grains.
		Principal quantum number, Chapter 2.
	$n_j(X_r)$	Particle density of atoms of element X in stage r of ionization, level j of exitation.
	$n(X_r)$	Particle density of atoms of element X (or of molecules) in stage of ionization r.
	n_X	Particle density of element X in all stages of ionization.
	$n_X{}^*$	Value of n_X in "equivalent thermodynamic system."
	n_f	Value of n for field particles.
N		Number of particles in a column along the line of sight 1 cm² in cross section.
	$N_j(X_r)$	Number of atoms of element X in stage of ionization r in the line of sight per cm².
p		Pressure; p_R, p_G, and p_B, pressure due to cosmic rays, gas, and magnetic field, equations (5-32)–(5-34); p_m, maximum external pressure at which an isothermal sphere is in equilibrium, equation (6-24).
		Particle momentum, Section 1.2.
		Polarization of radiation in magnitudes, equation (3-11).
		Collision parameter; i.e., distance of closest approach in the absence of mutual forces.
P		Power per cm³; P_c, power lost by collisions among clouds; P_u, P_s, power from absorption of ultraviolet radiation and from supernova shells.
	$P(v) dv$	Fraction of particles whose radial velocity lies within the range dv.
q		Stokes parameter in magnitudes, equation (3-14).

Q	Q_e	Efficiency factor for extinction by solid particles; Q_{eE}, Q_{eH}, values of Q_e for axis of spheroid parallel to **E**, **H**, respectively, in polarized radiation.
	Q_a	Efficiency factor for pure absorption by solid particles.
	Q_s	Efficiency factor for scattering.
r		Radius; distance from central star; r_i, r_s, radius of an ionization or shock front.
	r_S	Radius of H II region in radiative equilibrium.
R		Radius of a cloud. Subscript referring to energetic particles, including cosmic rays; i.e., U_R, p_R.
	R_m	Rotation measure, equation (2-71).
	R_V	Ratio of visual extinction to color excess, E_{B-V}.
s		Path length, especially along ray of electromagnetic radiation. Integrated value of s_v, equation (2-21). Subscript referring to solid material in interstellar grains; i.e., ρ_s, T_s.
	s_v	Atomic cross section for absorption of radiation of frequency v.
	s_u	Integrated value of s_v, uncorrected for stimulated emission, equation (2-25).
S		Surface; dS element of surface area.
	$S_u(r)$	Net ultraviolet flux in photons per sec flowing out through a shell of radius r; $S_u(0)$, ultraviolet luminosity of star in photons per sec.
t		Time in seconds.
	t_c	Self-collision time, equation (4-6).
	t_s	Slowing-down time, equation (4-7).
	t_E	Energy-loss time for electrons, equation (4-14).
	t_T	Cooling time, equation (4-69).
	t_f	Free-fall time for a cold uniform sphere, equation (6-30).
T		Temperature in degrees Kelvin; kinetic temperature of gas.
	T_b	Observed brightness temperature.
	T_c	Color temperature of stars for ultraviolet radiation.
	T_E	Temperature in equilibrium.
	T_s	Temperature of solid material in grains.
u		Stokes parameter in magnitudes, equation (3-15).
	u	Relative velocity in collisions between particles. Velocity of fluid relative to shock front or ionization field.
	u_R, u_D	Critical values of u for an ionization front, equations (5-43) and (5-44).

SYMBOLS

U		Potential energy of two particles.
		Energy density; U_R, U_{Rs}, energy density of cosmic rays, and of MeV protons produced by supernova shells.
	U_ν	Energy density for photons with energy between ν and $\nu + d\nu$, divided by $d\nu$.
v	**v**	Fluid or macroscopic velocity.
V		Volume; dV, volume element
		Phase velocity of a wave.
	V_A	Velocity of Alfvén wave, equation (5–10).
	V_i, V_s	Velocity of ionization or shock front.
w		Random velocity.
	w_m	Root mean square value of w.
	w_D	Drift velocity of grains through gas or of ions through neutral atoms in quasi-steady state, equations (5-91) and (6-50).
W		Equivalent width of an absorption line in frequency units; W_λ, equivalent width in wavelength units.
		Dilution factor, equation (4-37).
x		Ratio of grain circumference to wavelength, Chapter 3.
		Fraction of H atoms ionized, Chapter 4.
Y		Relative abundance of helium by mass.
Z		Ion charge in units of the proton charge.
α		Recombination coefficient; α_j, $\alpha^{(r)}$, recombination coefficients for level j and for all levels with $n \geq r$.
β		Dispersion of Doppler shift frequencies times $2^{1/2}$, equation (2-32).
		Ratio of H ionization energy to kT, equation (4-58); β_c, value of β for T equal to color temperature of star.
		Parameter describing the free fall of a cloud, equations (6-28) and (6-29).
	β_{rjf}	Rate coefficient for photoelectric ionization of an atom in level j, in stage of ionization r.
	β^*_{rjf}	Value of β_{rjf} in "equivalent thermodynamic system."
γ		Damping constant in frequency units, Chapter 2.
		Power of E^{-1} in particle density of relativistic particles, equation (2-51).
		Ratio of specific heats, Chapter 5.
	γ_{jk}	Rate coefficient for excitation from level j to k by collisions.
Γ		Total gain of kinetic energy of interstellar gas, per cm³ per sec.

	$\Gamma_{\zeta\eta}$	Component of Γ due to collisions between particles of types ζ and η; subscripts e, i, H, m, and R denote electrons, ions, neutral H atoms, molecules, and cosmic rays, respectively.
δ		Alignment parameter for interstellar grains, equation (4-103).
ΔX		Increment of X.
ε		Efficiency; ε_u, ε_s efficiency of accelerating interstellar clouds by ultraviolet radiation and by supernova shells.
	ε_{ff}	Energy radiated per cm^3 per sec by free–free transitions.
ζ	ζ_{rjf}	Rate coefficient for collisional ionization of an atom in level j, in stage of ionization r.
η		Electrical resistivity in e.m.u., equal to 10^9 times the resistivity in Ohm-cm.
θ		Angle; θ_p, position angle of plane of vibration of polarized light, in galactic coordinates.
κ		Wave number, equal to $2\pi/\lambda$; $\varkappa$, vector wave number in the direction of propagation.
	κ_J	Critical wave number for gravitational instability on Jeans theory, equation (6-7).
	κ_ν	Absorption coefficient per cm^3 for radiation of frequency ν.
λ		Wavelength in cm.
Λ		Quantity whose natural logarithm appears in formula for ion–ion collisions, equation (4-13). Total loss of kinetic energy of interstellar gas per cm^3 per sec.
	$\Lambda_{\zeta\eta}$	Component of Λ due to collisions between particles of types ζ and η; subscripts e, i, p, H, and m denote electrons, ions, protons, neutral H atoms, and molecules, respectively.
μ		Mean mass of the gas per particle.
	μ_i	Mean mass of the gas per positive ion.
ν		Frequency in cycles per sec (Hz).
	ν_1	Frequency at Lyman limit in neutral H.
ξ		Sticking probability in a collision of an atom with a grain; ξ_a, value of ξ for neutral atoms.
ρ		Mass density in g cm^{-3}.
	ρ_s	Internal density within grain.
	$\rho_\mathrm{I}, \rho_\mathrm{II}$	Density in H I, H II regions.
σ		Cross section for collisions between particles.
	σ_{cj}	Cross section for capture of an electron in level j.
	σ_g	Geometrical cross section of a grain.
τ	τ_ν	Optical thickness at frequency ν, equation (2-2).

SYMBOLS

ϕ	$\phi(\mathbf{r})$	Gravitational potential as a function of position, $\mathbf{r}$.
	$\phi(\Delta\nu)$	Profile of absorption coefficient, equations (2-30) and (2-31).
	ϕ_1, ϕ_2	Functions of β occurring in equation (4-57) for α, $\alpha^{(2)}$ respectively.
Φ	$\Phi(x)$	Error function.
χ	χ_r	Ionization energy of atom in stage of ionization r.
	χ_1, χ_2	Functions of β occurring in Λ_{ep}, equation (4-75).
ψ	$\psi_m(x)$	Functions characterizing the mean photoelectron energy, $\overline{E_2}$, equation (4-74).
ω		Solid angle.
		Angular frequency $2\pi\nu$.
Ω		Angular velocity of rotation.
	$\Omega(j, k)$	Collision strength for transitions between levels j, k, equation (4-20).
∇X		Gradient of X.
$\nabla^2 X$		Laplacian of X.
$\langle X \rangle$		Mean value of X, especially the value averaged over a Maxwellian velocity distribution or over space.

Index

Absorption, 59; see also *Extinction, general*
Absorption coefficient, 10, 11, 21
Absorption cross section, 15, 17, 25
Absorption efficiency factor, 59, 143
Absorption lines, 36–44
Absorption profile function, 17; see also *Line profile*
Abundances, relative, of the elements, 48, 122
Accretion by clouds, 224
Adiabatic compression, 236
Albedo, 58, 71
Alfvén velocity, 164, 169, 172, 243
Alfvén waves, 164, 238, 243
Ambipolar diffusion, 238
Angular momentum, 230, 231, 234, 242
Axial contraction of cloud, 231, 234

Babinet's principle, 61
Balmer lines, 29, 118
Black-body radiation, 10
  universal radiation field, 108, 145
  ionizing radiation in H II regions, 118, 130
  radiative heating of grains, 144
Boltzmann equation, 12
Boltzmann factor, 32
Bremsstrahlung, 21; see also *Free–free radiation*
Bright nebulae, 81

Bright rims, 32, 203, 205
Brightness temperature, 11
  21-cm line, 28, 37
  H radio recombination lines, 35
  OH emission line, 36
  synchrotron radiation, 23, 25, 45

Ca atoms, absorption lines, 1, 39–43, 171
  ionization, 120–125
Cas A, 37
CH and CH^+ molecules, 44, 126, 157
Centrifugal force, 177, 230
Chapman-Jouguet hypothesis, 203
Chemical composition, 48, 86, 122
Cloud collisions, evaporation of grains, 154–155
  kinetic temperature and cooling time, 140
  rate of energy dissipation, 172
Cloudlets, 28
Clouds, 3
  21-cm data, 28, 37–38
  acceleration by rocket effect, 205–207
  assumed parameters, 170–171
  average number in line of sight, 40, 78–79, 84
  Balmer line data, 31
  Ca II absorption line data, 41
  color excess data, 82–84
  concentration of grains in, 212

Clouds (*continued*)
 density of Na I in, 43
 gravitational collapse, 221–224
 high-velocity, 29, 40, 181
 ionization of, 202–205
 ionization equilibrium in, 124
 See also *Fragmentation, High-velocity clouds, Large clouds,* and *Standard clouds*
Collapse of cloud, 220
Collision strength, 96–98
Collisional excitation and deexcitation, 31, 89
 electron collisions, 95–98
 H-atom collisions, 98–99
Collisions between clouds, 154, 182
 energy balance, 172
Collisions between particles, 88–101
 See also *Cross sections, collisional*
Color excess, 66
 correlation with polarization, 73
 galaxies, globular clusters, 67, 70
 statistical properties, 82–84
 See also *Extinction, selective*
Concentration factor, F, 31
 clouds of dust, 85
 ionized hydrogen in Galaxy, 34
 neutral hydrogen, 124
 Orion nebula, 31–32
Continuous absorption and emission by gas, 21–22
Cooling time, 127
 H I regions, 139
 H II regions, 135
Cosmic radiation, 4, 23, 47–49
 disruption of grains, 156–157
 effect on distribution of gas, 177–181
 energy density, 48, 196
 ionization by, 110–111, 119–120
 pressure produced by, 169–170
 production by supernovae, 196–197
 radiation by electrons, 44–47

Crab nebula, 33
Cross sections, collisional, 89
 electron capture, recombination, 99–100
 elastic collisions, 93–94
 inelastic collisions with electrons, 96–97
 inelastic collisions with H atoms, 98–99
 ionization of H by energetic particles, 101
Cross sections, inelastic, 95
Curve of growth, 17, 19, 21
Cyclotron frequency, 163, 208
Cygnus loop, 33, 201

D-type ionization front, 186
 D-critical fronts, 186, 203
Dark nebulae, 81
Density of interstellar matter, 7
 distribution in Galaxy, 175–183
 neutral H, 27
Detailed balancing, 95, 112, 131
Diffraction of light around grains, 61
Diffuse ultraviolet radiation, 114, 129, 131–132, 134, 188, 204
Diffusion time, 239–240
Dilution factor, 107, 144
Dissociation of molecules, 125
Dissociative recombination, 101, 240
Doppler broadening, 18, 36
Doublet ratio, 20, 41, 43
Drift velocity, 209, 239
Dynamical friction, 92

Efficiency of acceleration, H II regions, 193–194
 supernova shells, 201
Efficiency factors for grains, 58–59
 complex particles, 64–65
 cylinders, 62–63
 spheres, 59–61, 207

Einstein coefficients, 14, 16, 20, 24, 102; see also *Spontaneous radiative transition probability*
Electric charge on grains, 145, 146, 208
Electrical resistivity, η, 162
Electron density, n_e, 31, 34
 n_e in ionization equilibrium, 112–114, 120, 138
Electrons, relativistic, 4
 measured in cosmic rays, 47
 radio emission from, 45–46
Emission coefficient, 10; see also *Emissivity*
Emission lines, 24–33, 35–36
Emission measure, E_m, 30, 31, 33
Emissivity, 10, 11
 Balmer lines, 30
 free-free radiation, 21, 45
 line radiation, 14–15
 synchrotron radiation, 22
Energetic particles, 4, 47–49
 ionization by, 110–111, 119–120, 136, 182
 pressure produced by, 169–170
 production by supernovae, 196–197
 See also *Cosmic radiation*
Energy gain by gas, 127–129
 in H I regions, 136–137
 in H II regions, 129–132, 134
Energy levels, 12, 14, 16, 97–99, 100, 107–109
 in CN molecule, 44, 108
 in H atoms, 24, 35, 98
 in H_2 molecule, 98
 in O II atom, 31
 in O III atom, 32, 105, 108
Energy loss, 128
 in H I regions, 137–138
 in H II regions, 133–135
Energy-loss time, t_E, 95, 104
Entropy, 127

Equation of continuity, 162, 215, 228
Equation of radiative transfer, 10, 114, 188
Equilibrium temperature, 127
 in H I regions, 138–139
 in H II regions, 129, 131–132, 134–135
Equipartition of kinetic energy, 140–141
Equipartition time, 92, 199–200
Equivalent thermodynamic system, 13
Equivalent width, 18, 41–44
Escape of Lyman α photons, 118
Eulerian coordinates, 228
Evaporation of grains, 155
Excitation, 11–12, 106–109
Excitation energy, 97
 in CN molecule, 44
 in H atoms, 24
 in H_2 molecule, 98
 in O II atom, 31
 in O III atom, 105
Excitation temperature, 44
Extinction, general, 57–58, 69–71
 corrections for, 30, 33
 globular clusters and galaxies, 70
Extinction, selective, 2, 65–69, 82
 in clouds, 43, 84–85
 See also *Color excess*

Faraday rotation, nonthermal radiation, 53
 radio sources, 50–51, 77, 171
Ferromagnetic relaxation, 152
Field particles, 88, 92, 94
Forbidden lines, 30, 108
Fragmentation, 225, 244, 245
 magnetic, nonrotating cloud, 232–234
 nonrotating, nonmagnetic cloud, 225–229

Fragmentation (*continued*)
  rotating, nonmagnetic cloud, 230–231
Fraunhofer diffraction, 61
Free-bound emission, 21
Free-fall time, 226, 235
Free-free radiation (Bremsstrahlung), 21, 131, 135
Free-radical grains, 65, 69, 72, 76, 145

Galactic rotation, 1, 25–27, 29
  effect on equilibrium density distribution of gas, 177–178
  effect on gravitational instability of gas in Galaxy, 219–220
  kinetic energy, 175
Galactic thickness, $2H$, from 21-cm data, 27
  from Ca II absorption-line data, 42
  in Faraday rotation analysis, 52
  in galactic equilibrium analysis, 180
  from selective extinction data, 67
Galaxy, spherical, 176
Gaunt factor, 21, 34, 116
Grains, evolution, disruption by energetic particles, 156
  disruption by sputtering, 156
  evaporation in cloud collisions, 154–155
  growth, 152
  physical properties, composition, 64, 69, 72, 76
  electric charge, 145–147, 208
  internal temperature, 142–145
  mass density, 69
  orientation, 74, 147–152
  radius, 68, 74, 76
  See also *Graphite grains*, *Free-radical grains*
Graphite grains, 64, 69, 72, 75, 145, 152

Gravitational collapse, 221–224, 233, 235, 244
Gravitational energy, 165, 221, 236
Gravitational instability, 214–224
Gyration frequency, 163, 208

H I regions, 3
  21-cm data, 24–29
  ambipolar diffusion, 238–240
  acceleration of H I cloud by rocket effect, 205–207
  acceleration of H I shell, 189–193
  boundary with H II regions, 113–114
  deviations from Maxwellian distribution, 105–106
  drift velocity of grains, 210
  equipartition of kinetic energy, 140–141
  ionization equilibrium of Na, Ca, 123–124
  ionization of H, 119
  isothermal sound speed, 170
  kinetic temperature, 136–140, 171
  ratio of dust to gas, 72
H II regions, 3
  density fluctuations, 31–32
  destruction of H_2 molecule, 125
  deviations from Maxwellian distribution, 105
  drift velocity of grains, 209
  expansion around young star, 183, 187–194
  ionization equilibrium of Na, Ca, 122–123
  isothermal sound speed, 170
  kinetic temperature, no impurities, 129–132
    with impurities, 133–135, 171
  ratio of dust to gas, 72
  Strömgren radius, r_S, 114–117
  thermal radio emission, r.m.s. n_e, 33–35

INDEX

H_2 molecules, 125–126, 158, 170
Helium, effect on Strömgren radius, r_S, 119
  effect on temperature of H II region, 135
Helium–hydrogen ratio, 170, 184, 199, 209, 210, 237, 240
High-velocity clouds, 29, 40, 181
Hydrogen 21-cm line, 3, 24
  in absorption, 37–38
  in emission, 25–29
  excitation, 109

Index of refraction, m, 59, 60, 62, 64, 68, 74, 75
Intensity, specific, 10
Intercloud medium, 42, 182, 221
Ionization, 2
  hydrogen, in H I region, 119–120
  in H II region, 113
  sodium and calcium, 120–125
  in steady state, 110
  stage of, 12
  in thermodynamic equilibrium, 11
  See also *Stage of ionization*
Ionization front, 183–187
  D-critical front in a dense cloud, 203–205
  initial R-type, 187
  later D-type, 189–192
Ionization probability, 110–111, 120–121
Isothermal shock, 168, 169, 184, 200

Jeans length, 216, 226

Kelvin-Helmholtz contraction, 235
Kinetic energy, 165, 174, 196, 198, 201, 233
Kinetic temperature, 2, 38
  from 21-cm data, 28
  adopted values, H I, H II, 171
  from emission-line ratios, 32
  equilibrium value, 126–142
  from H radio recombination lines, 35
Kirchhoff's law, 11, 21, 100

Lagrangian coordinates, 228
Large clouds, 81, 83, 224
Line absorption and emission, 14
Line profile, 17, 29, 37, 41, 82
Linearized perturbation equations, 215
Liouville's theorem, 4
Lyman limit, 115, 118, 173, 187
Lyman lines, 31, 118

Macroscopic equations, 163
Magnetic energy, 165
  in collapsing cloud, 222–223, 236
  energy density, 178
Magnetic field, Alfvén waves, 164
  dynamical equations, 162
  shock fronts, 169
  synchrotron emission, 22–24
Magnetic field in Galaxy, 4
  21-cm line, 39
  decrease in magnetic flux, 238–242
  Faraday rotation of radio sources, 52
  field strength adopted, 170
  fragmentation, 232–234
  gas distribution in Galaxy, 178–181
  gravitational collapse of cloud, 222–223
  motion of a grain, 208
  nonthermal emission from Galaxy, 50
  OH lines, 36
  optical polarization, 77–79, 151–152
  particle motion in supernova, 196–197
Magnetic susceptibility, 150

Marginal stability, 217
Maser amplification, 17, 36, 39
Maxwellian velocity distribution, 18, 89, 90, 103, 106, 128, 146
Microscopic equations, 163
Molecular H_2, 125–126, 158, 170
Molecular lines, CH, CH^+, CN, 44
OH, 35–36, 38
Molecule formation, 125–126, 157–158

Na atoms, absorption lines, 39–43
  ionization, 120–125
Nonthermal radio emission, 45–47
  plane polarization, 51–53

OH molecules, 35–36, 38–39, 126
Ohmic dissipation, 163, 238, 244
Oort limit, 7
Optical thickness, 10
  21-cm line absorption, 25, 37
  in contracting cloud, 236–238
  extinction by grains, 71
  free-free radio wave absorption, 45–46
  for ionizing radiation in H I regions, 113, 184
  line absorption, 17
Orientation of grains, 73–74, 147–152
Orion arm, 27, 37, 53, 77
Orion nebula, comparison of theoretical and observed line intensities, 102
  electron density, concentration factor, 31–32, 34
  emission measure, 30
  kinetic temperature and velocities, 32
  radius and electron density, 81
  spectrum of thermal radio emission, 33
Oscillator strength, 16, 24, 96
Overstability, 217

Paramagnetic relaxation, 151
Partition function, 12, 13, 15, 112
Perseus arm, 27
Phase function for scattering by grains, 58, 71–72
Photoelectric ionization, 110, 120, 127
Planck function, 10, 118, 144
  See also *Black-body radiation*
Plane of vibration, 51, 73, 74
Plasma frequency, 22
Pleiades, 81
Poisson distribution, 82
Poisson's equation, 6, 162, 216
Polarization, circular, 21-cm absorption lines, 39
  OH emission lines, 36
Polarization, plane, infinite cylinders, 64
  galactic radio emission, 53
  light from reddened stars, 72–79
  nonthermal radio sources, 50–51
  synchrotron radiation, 23
Positrons, 48
Pressure equilibrium, 181–182
Protocluster, 235, 244
Protostar, 231, 234

R-type ionization front, 185, 187–188
Radiation damping, 20
Radiation energy density, 14, 102, 107
  H II regions, 114, 130
  incident on grains, 143–144, 211
Radiation pressure, 202
  cloud in pressure equilibrium, 237
  grains near a hot star, 207–210
  grains subject to galactic light, 210–212
Radiative attraction between grains, 211–212
Radiative capture, hydrogen, 99–100, 110, 115–117

INDEX

nonhydrogenic atoms, 120
  See also *Recombination coefficient*
Radiative ionization probability, 110, 112, 120
Radio continuum emission, 33
  See also *Nonthermal radio emission*, *Free–free radiation*, and *Faraday rotation*
Radius of gyration, 48, 163, 195
Radius of H II region, 116–117, 119
  expanding region, 192
  initial ionization, 187
Range of 2 MeV protons, 195
Rate coefficient, 89, 102
  collisions with atoms, 99
  collisions with electrons, 98
  radiative ionization, 110
Rayleigh-Gans particles, 75
Rayleigh scattering, 60
Recombination coefficient, 100, 110
  hydrogenic atoms, 116–117, 174, 204
  Na, Ca, and C atoms, 120
Recombination lines of H, 29–31, 35
Reflection nebulae, 72
Relativistic particles, 47–48, 165, 177
  electrons, 22, 47
Rocket effect, 205–207
Rotating cloud, 230–231
Rotational energy of Galaxy, 175
Rotation measure, R_m, 51–52, 53
  See also *Faraday rotation*

Saha equation, 12–13, 14
  See also *Ionization*
Saturation of absorption lines, 19, 42
  See also *Self-absorption*
Scattered galactic light, 71–72
Scattering by grains, 58–59, 71
Selective extinction, see *Extinction, selective*

Self-absorption, 25, 27
  See also *Saturation of absorption lines*
Self-collision time, t_c, 91, 93, 95, 106
Shock fronts, 166–170, 172
  expanding H II regions, 183, 189–192
  gravitational collapse, 224, 236, 238
  H I cloud, 223
  supernova shells, 195–199, 201
Slowing-down time, t_s, 91, 93, 94
  equipartition of kinetic energy, 140
  grains moving through gas, 208
  ions moving through gas, 239
Snowplow model, 200
Sodium absorption lines, 39–43
  See also *Na atoms*
Solar wind, 49
Sound velocity, 33, 162, 164, 167
  expanding H II regions, 189, 190
  H I and H II regions, 170, 172
  ionization front and isothermal shock, 184–185
  rocket effect, 207
Specific intensity, 23
Spheroids, dielectric, 63, 74–75, 80
Spiral arms, 3, 27, 50, 220
  See also *Orion arm* and *Perseus arm*
Spontaneous radiative transition probability, 14, 16
  21-cm line, 24
  effect on relative population, 107–109
  forbidden lines in visual, 97
  H radio recombination lines, 35
  OH radio lines, 35
Sputtering, 156
Stage of ionization, 12, 14, 109–110, 120

Standard clouds, 83–84, 167
Statistical weight, 12, 24, 97, 98–99
Steady state, electric charge on grains, 145–146
  excitation, 102
  ionization, 110
  kinetic temperature, 126–127
  temperature in grains, 142, 143
Stimulated emission, 15, 16, 17, 19
Stokes parameters, 76–78
Strömgren radius, 116
  See also *Radius of H II region*
Supernovae, acceleration of grains by radiation pressure, 210
  expansion of shells, 194–202
  flux of energetic protons, 49, 111
  kinetic energy released, 174
  radio emission from remnants, 50
  star formation, 245
  velocity of Tycho's supernova, 33
Suprathermal particles, 4, 47
  disruption of grains, 143, 156
  heating of H gas, 136
  ionization of H, 110–111
  shock fronts, 169–170
  See also *Cosmic radiation*
Synchrotron emission, 22–24, 44–46, 50, 181

Temperature; see *Kinetic temperature*
Test particle, 88, 94

Thermal instability, 141–142, 222
Thermodynamic equilibrium, detailed balancing, 95–96, 112
  deviations, maser action, 17
  deviations in steady state, 31, 103–109
  emission and absorption of radiation, 10
  excitation and ionization, 12–13
  excitation in steady state, 31
Thickness of Galaxy; see *Galactic thickness*
Turbulence, pressure, 163, 178, 179
  velocity, 29, 32
Tycho's supernova, 33

Ultraviolet ionizing photons, 112, 116, 118, 184, 187, 203
Unidentified absorption lines, 44
Universal radiation field, 108, 1 45

Velocity dispersion in gas, 17–18
  21-cm line, 28–29, 37–38
  H radio recombination lines, 35
  K line, 40, 41
  visual emission lines, 32–33
Virial theorem, 164, 221–223
Visual absorption lines, 39–44, 52

Zeeman effect, 36, 39, 50, 171